THE SECRET LIFE OF WHALES

MICHELINE JENNER AM is a marine biologist and co-founder of the Centre for Whale Research (Western Australia) Inc. She has studied humpback and blue whales and conducted biodiversity and whale, dolphin and porpoise surveys since 1990. Micheline and her husband Curt Jenner AM received a Lowell Thomas Award from The Explorers Club for their work protecting blue whales in the Perth Canyon, Western Australia. In 2017, they were awarded the Australian Geographic Society Lifetime of Conservation Award. Micheline is also a Master Mariner with multiple captain's qualifications.

THE SECRET LIFE OF WHALES

MICHELINE JENNER

NEWSOUTH

A NewSouth book

Published by
NewSouth Publishing
University of New South Wales Press Ltd
University of New South Wales
Sydney NSW 2052
AUSTRALIA
newsouthpublishing.com

First published 2017

10 9 8 7 6 5 4 3 2 1

National Library of Australia
Cataloguing-in-Publication entry

Creator: Jenner, Micheline, author.
Title: The Secret Life of Whales / Micheline Jenner.
ISBN: 9781742235547 (paperback)
9781742244037 (ebook)
9781742248448 (ePDF)
Subjects: Jenner, Micheline.
Marine biologists – Australia – Anecdotes.
Whales – Identification.
Whales – Ecology.
Whales – Behaviour.
Whales – Breeding – Western Australia – Kimberley Coast.

Design Josephine Pajor-Markus
Cover design Luke Causby, Blue Cork
Cover images Wayne Osborn
Illustrations Micheline Jenner

CONTENTS

To my dearest mum I dedicate this book.

Among the many things I have to thank you for,
I especially appreciate your encouragement
to adventure and explore the world.
Thank you.

The whales are absolutely incredible — they are so large and yet so graceful, so mellow, so stroppy, so quiet and yet so splashy and loud. Not just a hunk of blubber the size of a bus but a living, moving, thinking creature with order and organisation (even as transient as it may seem to us) and understanding. I give them lots of credit that they know what's going on.

Letter home from Hervey Bay, Queensland, May 1988.

Our task, those of us who care about cetaceans, is to make a place for cetaceans in the sea — to ensure what is rightfully theirs and has been for millennia — is well protected.

Erich Hoyt, *Marine Protected Areas For Whales, Dolphins and Porpoises: A World Handbook for Cetacean Habitat Conservation and Planning*, 2005.

INTRODUCTION

I have had the privilege of spending more than half my life living, breathing and thinking about whales. When I first went to Hawaii in 1987, employed on a J1 working visa as a whale researcher at the Pacific Whale Foundation (PWF) in Maui, I pinched myself every day. Fresh out of university after finishing my Master of Science degree in marine biology at the University of Auckland, I was so excited to wake each morning with the smell of frangipani flowers wafting in from the tropical garden, knowing I was actually living and working *in Hawaii*. I had gazed at the *National Geographic* topography map of the Hawaiian Islands on my bedroom wall for most of my childhood — and here I was! Moving back to Australia to research humpback whales off the remote north-west coast in 1990 was no less exciting.

My favourite day involves working with whales. After spending 12 hours (or more!) around whales, I love flopping into bed, absolutely dog-tired. As I fall into sleep, the day's wonderful sights and sounds fill my dreamy thoughts. I glide through blue water with whales … humpback whales, blue whales, dwarf minke whales — all different species of whales. The abundance of fresh salty air, warm sunshine, not to mention my arms and legs mildly aching from hours of 'photography yoga' (as I call holding multiple difficult positions for hours while taking photos of whales), all contribute to the thrill of whale research.

In 1987, I met my husband, Curt, a Canadian who was also part of the PWF research team running their internship program in Maui. A year later in a Tahitian lagoon I agreed to Curt's wedding proposal; I knew then our life would be one big adventure after another, and so far it *really* has been.

After we married on a yacht in Maui, surrounded by family and close friends, we volunteered with Ken Balcomb III's Orca Survey in the San Juan Islands in the north-west Pacific. Here the seed was sown to begin our own whale research project Down Under. Curt studied the nautical chart of the Western Australian coast and imagined where he would go if he was a humpback whale. The Dampier Archipelago was a similar latitude to the breeding and calving grounds of the Hawaiian Islands, so this seemed a good place to start.

In September 1989 Curt and I wrote to John Bannister (JB, as we affectionately know him), then director of the Western Australian Museum. We said that we were moving to Australia to undertake humpback whale research and would like to work with him. JB checked us out and met with us in Seattle in April 1990. Luckily for us, he decided to support these two blow-ins with enormous logistical assistance and invaluable friendship. Our journey has proceeded together over the last 27 years as we have conducted research collaboratively. Thanks, JB!

On 25 July 1990, we began our research in the Pilbara region. Our base was a research station owned by the Department of Conservation and Land Management (CALM) on Enderby Island, part of the Dampier Archipelago off the north-west coast of Western Australia. The station, the only shack on this A-class reserve, was used several times a year by staff scientists assessing the rare rock wallaby population. At JB's request, the directors of CALM kindly made the station available to us for five months during winter and spring.

Situated between two beaches on a narrow spinifex-covered isthmus on the eastern end of the island, the research station consisted of a 3-metres by 7-metres cyclone-proof tin shed covered on three sides by a pergola under which we slept, ate and worked. We ran a generator for four to five hours each evening to power lights, two fridges and our computer. The rocky, barren landscape spurred our imagination and we thought we had reached the end-of-the-earth or Enderbyearth!

During that first week in July 1990 we set up camp and ventured offshore in *Nova*, our 5.3-metre inflatable boat, as often as the weather allowed. On our first few forays we didn't see a single whale. Then, on 5 August 1990, we saw two humpbacks heading directly towards us. I positioned the boat behind and to the side of the pod and measured their pace as Curt balanced in a wind-surfing harness in the bow and began firing off the photographs. We had hardly begun excitedly identifying them (left and right dorsal and fluke photographs) when we spotted another pod of two right behind them. We immediately dubbed the area the 'humpback highway', and those whales were the beginning of our photo-ID catalogue, which now contains over 4000 whales.

After a couple of seasons in the Dampier Archipelago, we noted that the calves we encountered in this area were four to six to eight weeks old, and thus had to have already been travelling south for a month by whale tail. In June–July 1992 Curt and I spent a month at the Montebello Islands (70 nautical miles to the west of Dampier and 70 nautical miles north of Onslow), where we observed 100 humpback whales travelling on courses between 25 to 30 degrees and heading north-east towards the Kimberley. All roads led to the Kimberley: this was the whales' destination and we needed to get there to document this — somehow. This discovery led to the construction of our first water-based home, the sailing catamaran *WhaleSong*, and we began the next important stage of our research there in the mid-1990s. Concentrations of new-born calves in the silty waters of the Kimberley indicated we had finally found their calving ground. This discovery was not without considerable effort, but it was certainly worth it.

The Dampier Archipelago Humpback Whale Project was our first research work in Western Australia and would become the cornerstone research project for the Centre for Whale Research. The Centre was incorporated in 1993 so that we could accept tax-deductible donations — a necessary part of grassroots research on any topic. Since then we have diversified our studies to include several areas along the Western Australian coast and several different species of cetacea.

In those early days, we had to turn our hands to all sorts of extra jobs to fund our work. At the end of that first season in 1990, as we made our way from Enderby Island to Karratha and while dealing with the culture shock of having people around us again, we were plotting ways to generate income. We had to fund our second season! I hoped to get some illustration work, and Curt hoped for opportunities to work with a land surveyor. Eventually, we gained employment with a marine-based environmental consulting company. At the same time we would be writing reports and grant proposals for our whale work, preparing to print a range of black and white photographic greeting cards, and selecting a sketch to print onto t-shirts (some of our sidelines!) to sell for Christmas.

○

Having spent more than half our lives at sea conducting whale research, Curt and I have experienced big waves, wild lightning storms, tons of gorgeous sunrises and millions of fabulous sunsets. (I have thousands of sunset photos because each one is *unique and special*, as you know!) The midnight skies with dark, gloomy cloud cover or clear starlit vistas are our 'office' as much as the grey days or the delightful bright blue days. Our studies have taken us from coastal and tidally oriented regions to open ocean surveys circumnavigating Australia and travelling around west Africa. With four boats — the 5.3-metre inflatable, *Nova*, the 13-metre sailing catamaran, *WhaleSong*, the 24-metre expedition vessel, *WhaleSong II*, and our current steel-hulled 28-metre small ship, *Whale Song* — we have progressed from small inflatable boats to owning and operating a commercial ice-class vessel. We have joined the ranks of master mariners, gaining three commercial ship tickets each: Master Class 5 (AMSA), Master of Yachts 200T Limited (MCA/IYA) and Master of Yachts 200T Unlimited (MCA/IYA). The master mariner's creed is to preserve life and assist with preserving life, and in the course of saving whales we have also saved a few stranded sailors. At last count 22.

Our lives have been thoroughly enriched with the arrivals of our two girls, Micah and Tasmin. Having vagrant, high-seas parents may not suit children who want 'normal' lives. Fortunately, even in their early adult years they are gaining an appreciation of the slightly different upbringing we provided. Thanks M and T, you are my beach-combing buddies and snorkelling partners! Some of the ways our girls gained their life skills included learning to recognise and pronounce the names of channel markers at two years of age; mastering snorkelling with black-tip reef sharks at three; and reading out the names of boats on fellow yachties' boom covers at four. We did try to guide steadily, even though things were always very wobbly at sea! It seems, as they reach their late teens and early 20s, that all this sea-time might have been advantageous for our adventurous and boat-comfortable girls — both are pursuing maritime careers, because it is all so natural to them.

Thank you to our families for unreservedly supporting our wayward whale ways. We were determined, despite the odds, to 'make a go' of studying whales. To our friends who have also wholeheartedly backed us, thank you.

I hope you enjoy these snippets of some of our adventures with whales in Western Australia and beyond.

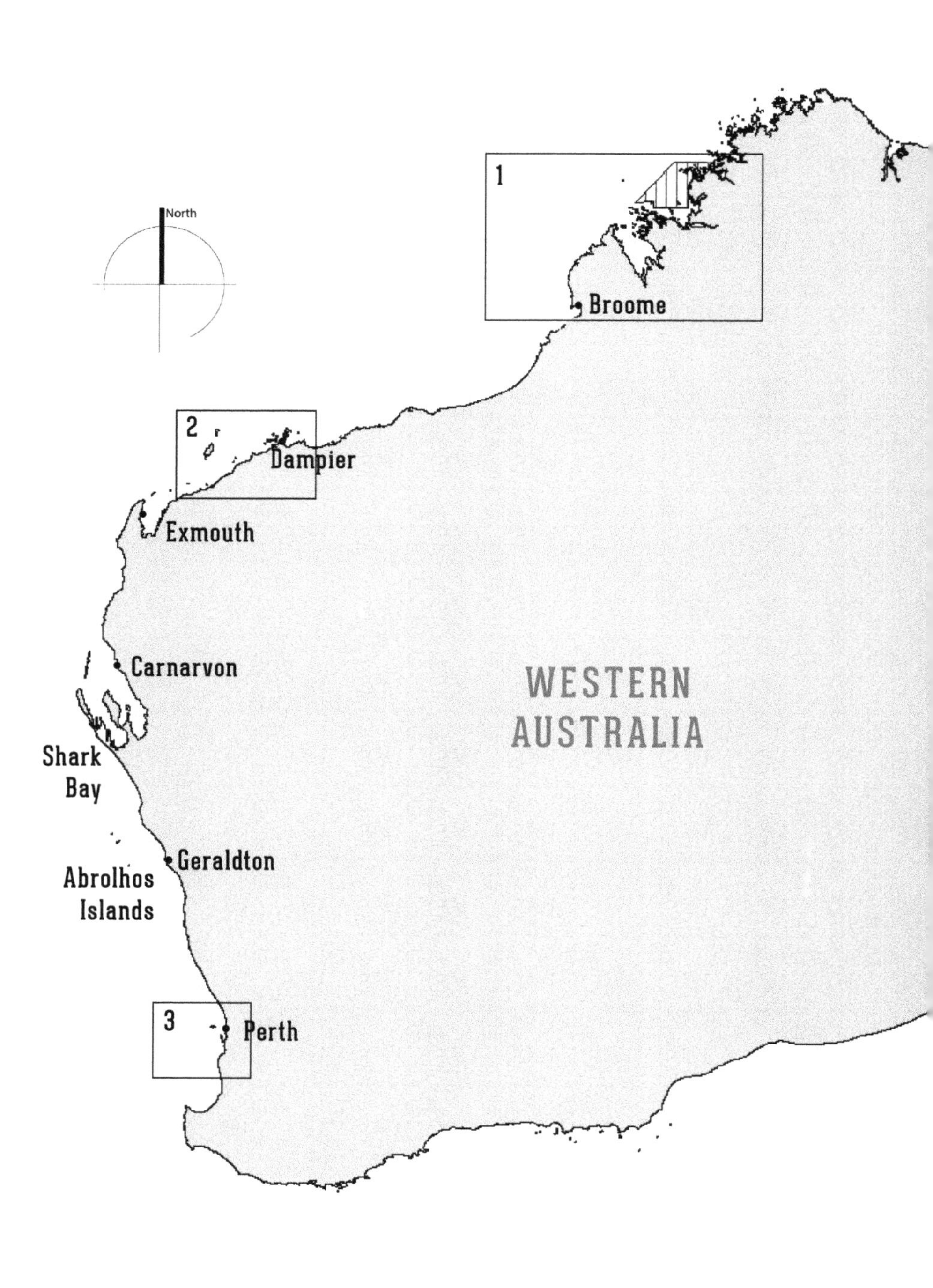
North
1
Broome
2
Dampier
Exmouth
Carnarvon
WESTERN
AUSTRALIA
Shark
Bay
Geraldton
Abrolhos
Islands
3
Perth

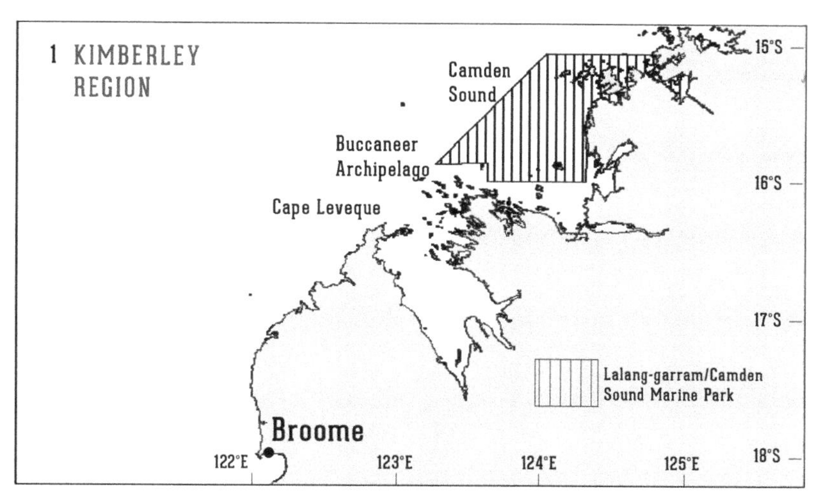
1 KIMBERLEY REGION
Camden Sound
Buccaneer Archipelago
Cape Leveque
Lalang-garram/Camden Sound Marine Park
Broome
122°E
123°E
124°E
125°E
15°S
16°S
17°S
18°S

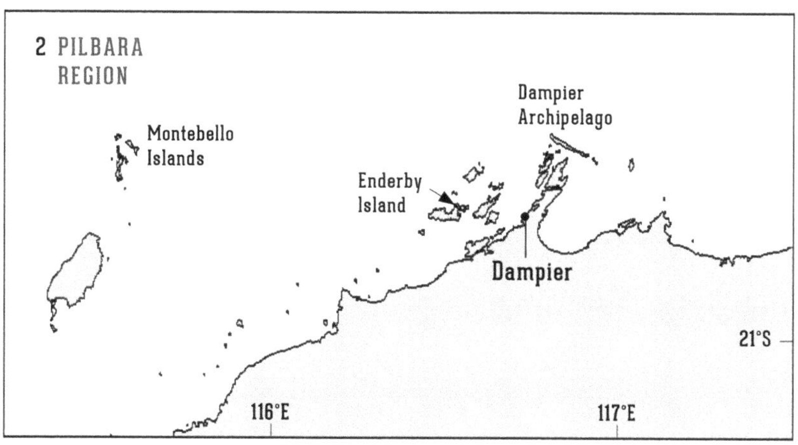
2 PILBARA REGION
Montebello Islands
Dampier Archipelago
Enderby Island
Dampier
21°S
116°E
117°E

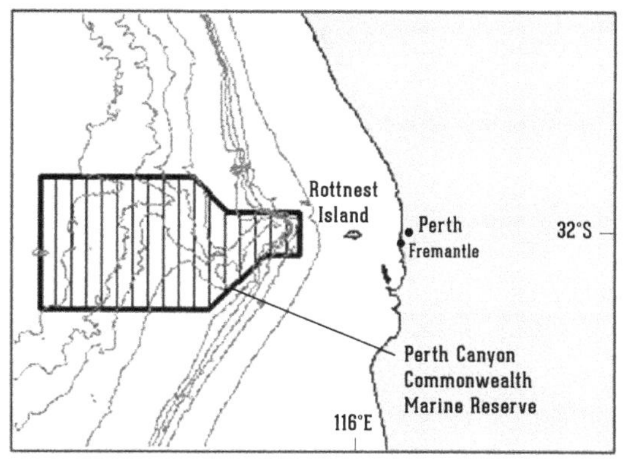
3 PERTH CANYON
Rottnest Island
Perth
Fremantle
32°S
Perth Canyon Commonwealth Marine Reserve
116°E

HUMPBACK WHALES

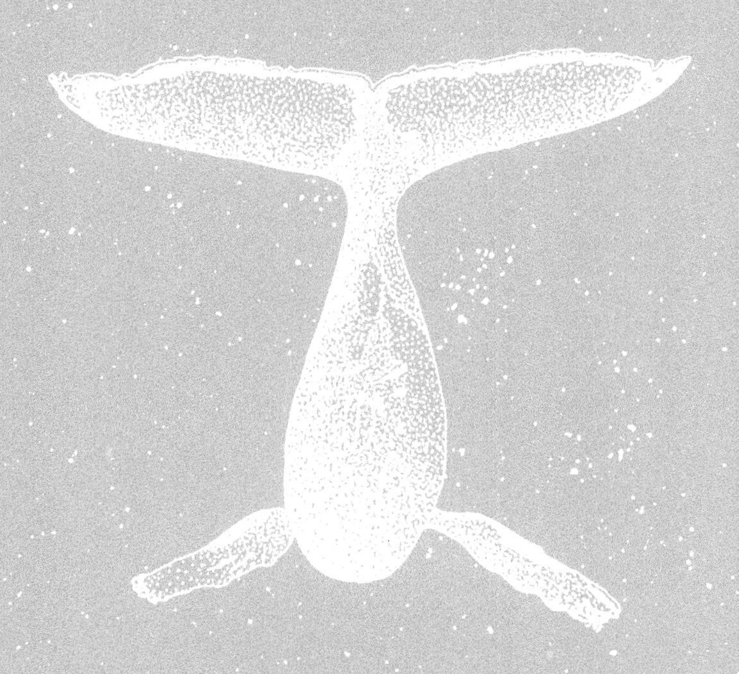

FIRST BREATH

'There's a calf there! It's brand new!'

It had been a typical day with our team of Centre for Whale Research assistants surveying for cetaceans (whales, dolphins and porpoises) in the Exmouth region off the coast of Western Australia, and Curt was steering our boat *WhaleSong II* towards the gap in the Ningaloo Reef at Tantabiddi. We were heading for our mooring inside the lagoon, near the Tantabiddi boat ramp on the western side of Exmouth Peninsula.

'Curt, there's an adult humpback over here, just north of the passage,' I said. Curt kindly listens. Right from the very first research projects that I joined in 1986 while I was still studying for my Master's degree, it became evident that my eyesight was better than I'd realised. When Curt and I met in Maui, I routinely found whales from afar, and thus our work in Australia was born on the back of visual surveys from small boats where I put my eyes to work. I got a huge kick out of finding whales and with great whoops and yahoos I would give excited directions to Curt as we skipped over the waves towards a blow from a whale.

As we approached this large, solitary humpback whale — what biologists call a pod or group of one — Curt smoothly reduced speed so I could collect the usual left and right lateral body photo-identification (photo-ID) images. These would be added to our steadily growing identification catalogue.

'This whale is huge. She's so wide, she's probably pregnant,' Curt observed as we motored quietly nearby.

Seeing a pregnant animal was not unusual. Pregnant whales heading north along Ningaloo Reef at this time of year are commonly sighted as they head for the Kimberley calving grounds over 1000 kilometres further north to give birth. As I began the process of photographing the left and right sides of this broad-backed whale, Curt quietly drifted *WhaleSong II*, our 24-metre expedition vessel, to within 100 metres of the whale. Fractiously, the large whale swam back and forth, first in a northerly direction, then to the south, then northward again, staying within a small area of calm water on the outside edge of the reef. As the whale made four surfacings parallel to us, I was able to get clear images of both sides of her broad lateral flank and dorsal fin. Next, she turned 90 degrees and headed towards the reef. She dived, lifting her tail fluke in a very high dive, nicely showing a perfect view of the underside of her tail. She went down for two minutes, and when she surfaced to breathe (in fact exhaling and inhaling), I saw a tiny pale grey calf bobbing right beside her.

'It's a calf!' I squealed. 'There's a calf there! It's brand new!'

Still drifting at a respectful distance (much further away than our research licences required), I took a whole lot of photos and then quickly looked at the images to see what was happening. Scrolling through, I enlarged a left-lateral-body photo-ID shot. The calf beside the mum's large flank was new all right — it was very little. This was a neonate. The photo showed a raspberry-red line rippling along the surface of the water just behind the mother's body. As she flexed her flukes swimming around her baby, the raspberry-hued blood tainted the water but quickly dispersed. The blood provided certainty that we *had* witnessed the birth of this little calf! Now we knew that not only did humpback whale cows and calves rest in this little spot by the reef on their southbound migration, but that on her northern migration this mother had chosen to give birth here as well.

I held my excitement and doggedly concentrated on holding the camera very still and in focus and just continued taking photos. 'Keep calm and carry on photographing' should be my motto. One of Curt's nicknames for me is 'lead-finger', not without reason. Quality is always better than quantity in any world, but with biological observations more is always better. High-quality digital images, readily available these days, can capture a multitude of nuances of behaviour and anatomy.

The whole crew positioned themselves on the bridge deck and flybridge and watched on with affection. Beside the mother this neonate tried to swim, but on this first effort it was struggling to take a breath and failing to stay afloat. The poor little thing was swimming around in tight, clockwise circles with only the tip of its rostrum (head) protruding. There was more sinking than swimming going on and it did not look good.

'There's something wrong with it,' Curt said, deeply concerned.

Fortunately, before we could process what was happening, the mother stepped in. I watched through my telephoto lens, my eyes opened wide. The cow dived below the surface and reappeared directly underneath the calf, her large dark form dwarfing her tiny pale grey charge. As the mother came to the surface, she slowly lifted the calf completely clear of the water. The calf was perched sideways right on top of its mother's flat upper jaw, near her blowholes, and was totally high and dry. The cow held the calf gently in this position for about 10 seconds, during which time the calf took what we believed to be its first breath.

As soon as the cow lowered her newborn back into the water, its little tail flukes began to beat like a child's wind-up toy being lowered into a bath. The calf flexed its wobbly tail flukes up and down like crazy and was off and racing. Before the cow could intervene, as if on autopilot the calf headed straight over to *WhaleSong II.* Just before colliding with us amidships, it turned to parallel our drifting vessel and swam half the length of us, much to the amazement and delight of all on board. We were looking right down on a brand-new calf swimming along our port side.

When it came to the surface, its little head popped out of the water on a 30-degree angle, making the water flow past the bumpy sensory tubercles on its flat upper jaw in endearing watery streams. The pale grey body was simply beautiful. It was tiny, only four metres long, and a just adorable replica of its 40 000-kilogram mother. As it

CONSCIOUS BREATHERS

Cetaceans — whales, dolphins and porpoises — are marine mammals, characterised by breathing air, giving birth and suckling their young.

Sleeping for cetaceans varies among species but mostly it's a short-term affair. While sleeping, a bottlenose dolphin shuts down half of its brain, going into what's known as an alpha state, and shuts the opposite eye. The other lobe of the brain stays awake at a low level of alertness. This alert side of the brain is used to watch for predators, obstacles and other animals, as well as signalling when to rise for air. Amazingly, after approximately two hours, the animal will reverse this process, resting the active side of the brain and waking the rested half. Humpback whales may sleep at the surface or just below for 10–15 minutes at a time, also engaging the alpha state to remain aware of their surroundings. To avoid drowning during sleep, it is critical that marine mammals retain control of their blowhole(s): the single nare (the cetacean equivalent of a nostril) present in odontocete whales, or the two nares in mysticete whales (such as humpbacks). The blowhole is a flap of skin that is thought to open and close under the voluntary control of the animal, meaning cetaceans possess a voluntary respiratory system. In order to breathe, a dolphin or whale must be conscious and alert to recognise that its blowhole is at the surface. This is known as conscious breathing.

Humans, of course, can breathe while the conscious mind is asleep; our subconscious mechanisms control this involuntary system. Equipped with a voluntary respiratory system, whales and dolphins must be alert at all times to trigger each breath. Whales change their breathing in relation to their activity level. A migrating humpback whale, for example, will have a three- to five-minute downtime, whereas a resting whale will breathe every 10 minutes or so. Experienced whale researchers can identify different species at great distances by the shapes of their blows and tell whether the whale is excited or resting, depending on the force of the blows.

opened its two blowholes, each the size of a pear, as wide as possible during the breathing process, it was clear this little tacker was on the move!

Its dorsal fin was completely folded over towards the right, as the fins of newborns are. In utero, the fin, which is mostly cartilage, is folded over. During the first few weeks it becomes upright, playing a stabilising role as the calf swims. Ness, one of our research assistants, took a video on her compact digital camera that we later played over and over. This neonate's first swim was accompanied by a delighted chorus of oohs and aahs from our research team.

After briefly checking us out, the little calf re-joined its mother. Together again, they moved back near the spot where the cow had given birth. Soon the calf would be learning the fine art of suckling. As we didn't want to disturb this lesson, once we were certain they were fine and doing well, Curt slowly drove *WhaleSong II* into the lagoon to moor.

At dinner that night we all talked excitedly about the beautiful surprise of the day. Had we *really* seen this? A calf being born! Wow! How lucky were we?

'We have to call it Tantabiddi!' I said.

'It was surprising to see how small and skinny that little whale was,' recalled Curt. We knew that humpback whale calves are born almost blubberless, but once you see them in the flesh, it's clear how vulnerable they actually are.

o

Most calves in the Western Australian humpback whale population are born in the warm waters off the Kimberley coast, near Camden Sound. During July and into August, the water temperatures are typically four or five degrees warmer than the sea surface temperature at Ningaloo Reef, where we had seen this little whale born. Scientists agree that this temperature makes a significant difference to the energy requirements for newborns and in some cases may assist their survival.

At that time, July 2009, this was the first documented birth of a humpback whale. Over the years several fishermen had recounted seeing various things, but none had captured photographs of the event. Our observations of the mother's pre-birth behaviour, her issuing blood and the newborn calf beside her, coupled with our detailed research notes and still and video images, made this sighting significant.

We had waited 20 years to witness this event. Most mammals give birth at night, and we suspected humpback whales followed this practice. Several years later, a research group in Madagascar documented the birth of a humpback whale calf while collecting photo-IDs from an active and combative pod. With several thousand researchers across the world studying humpback whales, we felt very privileged. This was a whale biologist's Holy Grail.

But why had this pregnant female given birth to her calf in such relatively cold water? Was she a very young mother who didn't know where to find warmer water? Did she leave her departure from the southern feeding grounds too late and simply run out of time to get all the way up the coast? Or had some human activities disturbed her progress? There had been an increasing amount of heavy offshore work in the North West Cape area north of Ningaloo. New oil and gas facilities had been installed on the edge of the continental shelf over the previous few years. We were concerned that noise from these facilities and the vessels servicing them could be disrupting the migration of pregnant animals northwards along the coast.

Female whales nearing their delivery time may well be sensitive to noise and unusual disturbances. At worst, this could result in their migration being delayed or even terminated so that they didn't travel past the source of the noise. These types of questions, and testing the many possible answers, keep the team at the Centre for Whale Research very busy — not only with humpback whales, but also blue whales, sperm whales and the numerous other whale and dolphin species that populate our coastline.

Can increased human activity in pristine areas coexist with recovering whale populations? Even though whaling has ceased, we don't know whether these complex giants can increase their numbers to the levels of a century ago. Has the world they live in, increasingly exploited by humans, changed too much? These are pressing questions that require urgent answers. In the last 50 years there has been significant development along the coastline of Western Australia. Iron ore from the inland Pilbara towns of Newman, Paraburdoo, Pannawonica and Tom Price is exported worldwide from shipping terminals at Port Hedland and Dampier — the busiest and heaviest tonnage ports in Australia — and transported right through the middle of the whales' migration routes. Iron ore and oil and gas from the seas beyond the continental shelf have been the mainstay of the Western Australian economy and — until the recent dip in resource prices — Australia's national GDP.

Following on from our original goal to monitor the growth of the humpback population from the Dampier Archipelago, the Centre for Whale Research undertook aerial surveys from a migration bottleneck at North West Cape, near Exmouth, from 2000 to 2009, and established that the Western Australian humpback whale population had steadily grown to between 33 000 and 36 000 animals. This was extremely exciting news. When whaling ceased in 1963, the population had been less than 500, and in the early 1990s their numbers were still only 2000 to 3000.

In a rare good news story for the environment, we have concluded that over the last 50 years, coastal and offshore development in north-western Australia has proceeded in parallel, and mostly in harmony, with the increase in the humpback whale population. Curt and I believe that a mixture of processes has been at work. That

Western Australia is so remote has been key. Low numbers of humans and few disruptive human activities during the critical period between the cessation of whaling and end of the 1990s allowed the whales unhindered use of the most favourable areas for rearing newborn calves. Now, through education and the efforts of groups like the Centre for Whale Research, good management practices have limited seismic surveys for petroleum in critical habitats, such as the calving grounds of the Kimberley, and reduced shipping in nursery areas like Exmouth Gulf. We feel fortunate to have been working at a time when this recovering population needed a helping hand. Curt and I were able to identify areas of coastline that were critical to the recovery of this population and then influence governments to protect them — a real privilege. For their own part, humpback whales appear to be a resilient and tolerant species. They doggedly go about their lives, migrating along coastal shores in close proximity to humans on all continents. They just need secure areas to breed and feed. Leaving them alone is the best protection we can give them.

A recent review of 13 global humpback whale populations has prompted the National Oceanic and Atmospheric Administration to take nine of these populations off the endangered list. Is this just a political move to enable funding to be passed on to more threatened coastal dolphin species? Or is it a true reflection of the relative health of these populations post-whaling? Regardless of the motive, long-term monitoring must continue to assess the Western Australian humpback whale population, and as that population grows, new strategies are needed. A new technique employed by the Centre for Whale Research, in collaboration with state and federal government agencies, is to utilise commercially available satellite imagery to count whales. Yes, we are counting whales from space! And it works. Curt analysed a swathe of north-south oriented images collected over two days in August 2016 that revealed 26 and 30 animals in Camden Sound. And four pods had calves with them! This technique will now become part of a regular monitoring program for this humpback whale population.

o

Tantabiddi and its mother had been passively subjected to an effective and globally standardised system of benign 'tagging' of individual humpback whales. We just quietly took photographs of them. Using high-resolution cameras, readily available these days, three images are taken of each whale. The digital photographs, managed in our custom-designed archive and matching system, are compared within and between seasons and locations, providing valuable long-term data on the life history and movements of individual whales as well as populations. The photographs are taken directly beside and behind each whale. We aim to collect left and right lateral body photographs, showing the dorsal fin and the scars and marks on the sides of the whale's flanks; and a photo of the beautiful patterns on the underside of the tail flukes. The black and white markings on the lower surface of each 5-metre tail fluke are as individual as our fingerprints and can be swapped like baseball cards between researchers looking for matches.

IT'S A MATING GAME

Most wild mammals give birth during the darkest hours, when their adrenalin levels are lower and their serotonin levels are higher. The cover of night is also an effective way to avoid predators. Humpback whales appear to follow this strategy, and the scarcity of birth observations, such as our encounter with Tantabiddi, would seem to confirm this. They are secretive, and this extends to mating activity, which has rarely been seen.

Southern right whales, by comparison, are quite different when it comes to mating. Their mating season is spent in the calm, protected bays along the south coast of Australia during the winter months. From a cliff-top at the Head of the Bight in the Great Australian Bight, these lovable rotund giants have been observed, photographed and studied while gathered in these protected south coast bays since 1991. Male southern right whales have been described as a 'life-support system for a set of testicles'. With a 3.6-metre penis and one-tonne testicles producing 4.5 litres of sperm, they engage in mating battles. Groups of males compete to inseminate females by literally flushing out the sperm of their competitors from their chosen females. However, their strategy may not be as successful as that of the humpbacks, since their population numbers are not increasing nearly so well. Southern right whales, particularly the mature females, show a strong tendency to return to the same breeding locations each year, so continued protection of this species and its coastal habitats is imperative.

When we first began work in the Dampier Archipelago in 1990, we aimed to collect as many photo-IDs as possible of migrating humpback whales. By matching the three images of individual whales year after year, it is possible to document a whale's life history. When we set up our research project on Enderby Island, many details of the humpback whale migration remained a mystery. By photographing every whale we encountered and matching the images of individuals within and between the seasons, we used a well-known statistical mark-recapture technique. The mathematical model takes into account the number of whales photographed and the number resighted within and between seasons and provides a robust understanding of population movement. If, for example, you wanted to assess the number of students in a school, you could take a sample of the school by taking a photograph of the students at the library each Tuesday at the same time. Mostly the same pupils would be present each Tuesday, so this would be looking on a fine scale at the movements of that class. If the students were photographed at the library each school day at the same time year after year, one could make a range of assumptions about changes in the population of the school and gather information on individuals as well. After our first four seasons based on Enderby Island, 30 years after the protection of humpback whales from whaling was introduced, we produced a humpback whale population estimate of between 2000 and 3000 individuals. The photos allow for modelling but alone they effectively tell the stories. A whale photographed by Wayne and Pam Osborn with

10 other animals near Rottnest Island in October 2009 was recognised from our July 1992 Montebello Islands expedition when Curt and I saw it migrating northward with another adult — this gap of 17 years remains our longest humpback whale resight!

As I watched Tantabiddi, this tiny newborn humpback calf who would spend almost 30 per cent of each day suckling, I was reminded of a previous encounter, also in the winter breeding season, with another pregnant humpback whale.

A SHAKY WHALE NURSERY

She was only a stationary shape, far in the distance on the Kimberley waters, when she caught my attention. It was a calm and warm August day in 1997, 12 years before Tantabiddi was born.

'Is that a boat? No, wait, let me check. It's flukes. Curt, can we go this way?'

From the bow of *WhaleSong*, our 13-metre custom-designed catamaran, with two-and-a-half year old Micah dancing beside me on the top deck, I rattled off the whale's compass bearing for Curt at the helm.

Changing course immediately, we made our way towards this humpback whale who was displaying a surface passive behaviour known as tail sailing or fluke extension. Surface passive behaviours (as opposed to active behaviours) include the regular surfacing movements of a humpback whale such as surfacing to breathe, rolling through the surface with the dorsal fin presented, surface lying and other typical non-active displays such as fluke-up dives and fluke extensions or tail sailing. The broad, beautifully shaped 5-metre flukes and instantly recognisable tail were motionless, held straight up in the air. The water gently lapped the narrow but muscular peduncle (where the tail joins the body) and as we slowly approached, the intermittent breaths of the tail's owner and glimpses of her genital area, told us that this was a female and, given her huge girth, there was no doubt she was pregnant. Was she near-term? Would we witness a wonderful event here and now? Wow!

On the deck of our catamaran *WhaleSong* there was ample space for all the crew to get a ringside seat. Being the pre-digital age, I took still photos with our Nikon F3, carefully checking the number of remaining shots on the slide film and quickly reloading another film when the female went down. Katie McCabe, our able research assistant, and our Earthwatch volunteers Pearle, Hilton and PC Mike, carefully documented the details of the encounter. Usually our research notes record the downtime of whales, from the moment the tail flukes go underwater until they surface next. The times varied depending whether the whale was migrating or resting. But this whale was different. She was so still that we recorded her uptime — that is, the time her flukes remained exposed at the surface.

The body of this almost 17-metre humpback whale was a uniform dark steel tone, almost black. However, in the silty turquoise water of the Kimberley, like most humpback whales, she appeared strangely brownish in colouration. As she slowly rose to the surface, the water ran across the curves of her broad rounded body in gentle wavy streams like a coating of glossy chocolate sauce. Shortly afterward, the slate grey skin returned to satin again. Her blows were laboured, strong and very slow. We saw the whole broad expanse of her body from her blowholes to her stabilising dorsal fin with every breath as she bobbed between sea and sky. With the last blow in a surfacing sequence, she exposed her blowholes, exhaled an explosion of air, inhaled instantly, and then rolled forward. Her round-shaped dorsal fin with the characteristic anterior hump for which these whales are named, peaked in a pointed arch. Next we saw her peduncle as she began lifting her flukes in slow motion. Expertly and carefully she held them stationary in the vertical position, just as they had been at our initial sighting.

After a few blows, dives and several 'surfacings' or uptimes with her tail extended out of the water, it appeared she was growing weary of this stance. When she repositioned herself a few metres away, after several blows her tail slid from the vertical to a more horizontal orientation, just above and parallel to the surface. At the end of the surfacing sequence it was interesting to note how she raised her tail flukes and just kept ever so slowly flopping them backwards without a single splash to rest the upper side on the surface of the water and exposing the lower surface to the air and the warm sun. Given the orientation of her flukes at the surface, we presumed her body curved in a large 'C' underwater, her head positioned beneath her flukes some 15 or so metres down in the murky water. Remaining still for so long save for minuscule drifting movements in the glassy conditions, the tropical Kimberley air dried and evaporated the water on her skin. The water lapped at the base of her tail, the tail stock, making the skin richly glossy

KEEPING COOL

Efficiently, the blubber of cetaceans keeps them warm. Like humans, their body temperature is 37 degrees Celsius. In fact, in warm water such as the tepid 27 degrees Celsius of the tropical breeding grounds, these animals must actively cool down. Holding their body's heat sinks, such as their highly vascularised tail flukes, out of the water allows the blood flowing through their veins to cool in the breeze; in a counter-current technique, this cooled blood returns to reduce the heat of the body's core. Tail-sailing for several minutes, even close to an hour sometimes, a humpback whale can effectively cool down its hot internal core. Sea lions also use their flippers and flukes as heat sinks.

in appearance — where the water was dry it became dull and almost a faded grey. Now her skin was burning; she was so immobile that she was actually at risk of getting sunburnt. The outer epidermal layers were drying such that the black skin could soon start to crinkle and crack ... Time passed and she lay motionless for 20 minutes at a time.

Then through the water a dark shape ascended towards the surface 15 metres from her.

'There's another whale here! Wow, listen, it's singing!'

At first I just couldn't believe my eyes as another adult whale surfaced beside her. Since this animal had remained underwater much longer than the tail-sailing female — up to three times longer — we hadn't seen it, although we had been with her for almost an hour. There were some special things going on here: not only did this female have a whale accompanying her as the birth neared, but that whale was singing …

The unmistakable low-frequency grumbling moans, mercurial down-sweeps and higher-pitched trills and shrieks of a male humpback filled our ears. We were listening to a private concert.

Humpback whales are renowned for producing the most complex repetetive yet changing songs in the animal kingdom. The song we were hearing is generally considered to be sung by males challenging other males in battles over females (considered dominance polygyny). A female might judge and be attracted to a suitor by the length of his song as an indicator of his overall fitness.

As the tail-sailing pregnant whale surfaced to breathe every 20 minutes or so before resuming her unusual fluke presentation, we estimated her length to be between 16 and 17 metres. She was at least several metres longer than our 13-metre catamaran and about a metre longer than her male accomplice. This is not unusual for humpback whales since they exhibit reverse sexual dimorphism, meaning that adult females are larger than males.

This whale was huge. Her enormous girth, extending from her two blowholes to her dorsal fin, made the centre of her back almost flat like the water surface. Her dark body was flat and broad — she was as wide as a 5-tonne truck. And this bursting-at-the-seams female, which appeared to be very close to giving birth, had a male companion. Was this new? In our previous 10 years of humpback whale field observations we certainly hadn't encountered this before.

Maybe this isn't commonly observed because whale births have only rarely been observed. Maybe this is the norm. On the boat we started brainstorming. Accepted wisdom says that an adult accompanying a female humpback whale with a calf, usually referred to as an escort, is a male. These males play no role in the care of the calf; they just want an opportunity to breed with a female that knows what she is doing. But in this case the calf was yet to be born, so what was going on?

In the world of humpback whales, a male accompanying a female is not the father of the calf, but a male choosing the company of a fertile female in order to successfully pass on his genes that breeding season. Male whales *do* sing on the breeding grounds to attract females, but was this singing animal, which we presumed was a male, getting in line early? Or was something different going on? Without genetic samples from both adults and the soon-to-be-born calf, we would never know if this was the father of her calf of that season. This seemed very unlikely, we reasoned, given the almost year-long gestation period. It was equally unlikely that this was a young male who had simply made a mistake and not realised the cow was pregnant; the length of his song indicated that he was mature and would have known better. Was he just endearing himself to this successful female for an opportunity to mate after she gave birth? Not discounting anything during the long afternoon, I even wondered if the singer might be a female ... That really would be different.

Observing this scene unfold was Mike Searle and his team from Storyteller Productions, who were making their award-winning documentary *Before It's Too Late: Whale-Song*. It was 1997; we had experienced many wonderful encounters in the Kimberley, but this one was very special.

By late afternoon, after three hours of continuous observation, it was with great disappointment we had to leave this near-term mother and her singing escort. As we moved slowly away, noting the time, latitude and longitude, and finished up our notes with details of the photo frames shot and sketches of the whales, I thought about the next few hours for this pair. Would she give birth that night, as most mammals do, to the dulcet tones of her suitor? As we motored away, I looked back and saw the female about a mile away, continuing to lie at the surface. Where was the male? Was he still singing? Their secrets were not meant to be shared that day.

o

What was certain was the information on *WhaleSong*'s fuel gauge. The plan for the day had been to purchase fuel from Cockatoo Island, one of the islands of the Buccaneer Archipelago. Wishing the mother-to-be the very best for the birth, we made our course for Cockatoo. The Kimberley has 10-metre tides — the second highest in the world after Canada's Bay of Fundy — and its strong currents are unforgiving of human foibles. Getting caught out was never on our agenda. At the beginning of every day we always planned where we would anchor that night, but despite trying very hard not to be too delayed, whales sometimes distracted us. At times like this we usually reached for Plan 53B! Whales were the focus of our attention but working in the Kimberley you had to keep your wits about you, and balancing the two was always tricky.

Yampi Sound is bordered by the Australian mainland to the south and the islands of the Buccaneer Archipelago to the north-west. As we entered its swirling waters we were surrounded by beautiful ancient islands, and we continued to go over our wonderful whale encounter. I thought it was so sweet of the escort to be singing to the female. Even though his intentions may have been quite self-serving …

Without warning the boat rattled as though we were driving over corrugated iron. Curt, Katie and I all felt it and hurriedly looked for the cause. We examined the sails, the mast and the rigging. Nothing wrong there. Was there a line in the water? Had we picked up a float? Failing to find any fault with the boat, we looked around us at the islands for some clue. And there was our answer: rock-falls on Cockatoo Island, rocks tumbling down the face of the Piccaninnies, and full-scale landslides on Koolan Island. Immediately we figured it must have been an earthquake. Where was the epicentre and how had the pregnant whale fared? Was a tidal wave on the way?

As the tangerine light faded from the sky we anchored *WhaleSong* in the bay on the south side of Cockatoo Island and launched a shore party for fuel in our runabout *Mega*. Somehow, and perhaps kindly, I was elected to stay with the runabout while everyone got to lug fuel cans. Unfamiliar with the bathymetry, that is, the underwater features of the bay, I held the boat in thigh-deep water to prevent it from beaching on the rubbly shoreline in the quickly dropping tide. With 10 metres of tidal range, if you blinked or got distracted, you could easily get stuck, literally. *Mega* was a heavy 5.8-metre fibreglass boat, and with the water moving so rapidly, I struggled to keep it afloat and had to keep manoeuvring it offshore. It was well after sunset and the sky began to turn a dark orange. A few lights from the hostel and the refurbished mine camp on the shore were my only company. Looking to the north, the sky was even darker. To the west the remnants of the day were only an orange glow against an inky celestial expanse. Spookily silhouetted, a wooden jetty with extremely long piles designed for the big tides dominated my vista. A lone fisherman cast a hopeful line.

Soon I started shaking. What could possibly be wrong? I asked myself. The water was a warm 27 degrees Celsius and the air temperature was pleasant too. Surely I wasn't getting cold? Just get it together, I told myself. Well, yes, there were a couple of things that might have triggered my fear. We had just had an earthquake, right? Would there be another? Would I witness the demise of the fisherman when the old wooden structure crumpled beneath him? And then there was that little thought about crocodiles. *Crocodiles!* Yes, this is the Kimberley, home to ancient crocodiles swimming among ancient islands.

The dark sky, the dark beach, the dark water and little me struggling to keep the boat afloat were all bad enough. The possibility of crocodiles lurking in that dark soup had me searching the virtual blackout conditions for any sign. At best I could lose a limb; at worst, my life. No wonder I was a wreck!

To my immense relief the others eventually returned,

having been delayed by helpful and curious islanders, and we pushed the boat out even further into the terrifyingly dark Yampi Sound. I had survived.

Through my fear I had tried to concentrate on the female. Back on board *WhaleSong* we turned on the radio and heard that indeed there had been an earthquake — 6.3 on the Richter scale, the second largest since the 1989 Newcastle quake of 5.6. Alarmingly the epicentre was only 40 nautical miles from our position in Yampi Sound where we had felt the water rippling. The mother-to-be and the crooner must have been closer to the action. Had the earthquake moved the baby along? Was it too stressful for mum and baby? Had the calf survived? Was something wrong with the calf? Was that why the male was with her? I desperately hoped all was OK but of course I would never know … unless down the track we resighted her dorsal fin and fluke in our humpback whale photo-ID catalogue. How long would the male stay with her? Would she mate with him in a post-partum oestrus, or would she concentrate on that calf? Ah, all these questions!

PEC FIN RIDE

'Curt, there's another pod over here, it's quite active. Just head towards Bumpus Island and all those splashes.'

Fast-forward almost 20 years from our first work in the Kimberley and we are back traversing Camden Sound on board our current *Whale Song*. Unusually, the pectoral fin of another humpback whale is about to take centre stage. Humpback whales possess the longest pectoral fins of all whales and dolphins. At 5 metres in length and 1 metre at the widest point, they have the same bone structure as our arms (radius, ulna and humerus), our hands (carpal and metacarpal bones) and fingers (phalanges). Lumpy tubercles the size of oranges decorate the leading edge, with two larger tubercles roughly located at the first third and second third positions along the fin. Models of these raised protuberances have shown that they provide lift while the whale swims, and engineers are now copying them to design the blades of wind turbines. The scientific name of humpback whales is *Megaptera novaeangliae*, which translates from the Latin as 'big-winged New Englanders'. The 'wings' are their long pectoral fins, which they use to fly through the oceans on some of the longest migrations of any animal in the world. The New England reference relates to their prevalence on the north-east coast of the United States of America. Humpback whales really do fly.

This group of humpback whales was splashing and crashing in the characteristically turquoise water of the tropical Kimberley. Immediately we saw what all the fuss was about: three adult male humpback whales were focused on a newborn calf and its mother. The competitive males swam around and around her, getting closer and closer with every pass. Through the glassy calm water we could see the tiny humpback whale, less than two weeks old, its dorsal fin not yet fully unfolded and upright after the 10 to 11 months spent in utero.

With the males jostling for position to be the female's primary escort, at times the calf rested on its mother's back, totally out of the water. Then, desperately trying to protect her young calf, the mother lifted junior up and held it on the top of her pectoral fin, right where the shoulder joined her body. Her oar-like fin was extended perpendicular to her body and slightly curved underwater. The tip broke the surface in white-water sprays as she swirled away from the males in evasive manoeuvres to left and right. The calf, positioned parallel to its mother's enormous 17-metre body, was nestled in a very safe place, right in the crook of mum's arm. With mum's protective 'pec-fin carry' and 'body-block', the calf rode slipstream in the water out of harm's way. This was maternal love in action. She was determined to protect her little calf. Having invested a year

CHILD'S PLAY

As we've recorded the details of each whale sighting in the Kimberley's Camden Sound, we've observed a range of surface passive behaviours, particularly of calves resting beside protective mothers. They lie at the surface like undulating brown islands floating in the green water — the mother, a huge piece of real estate, the calf naturally much smaller. Often in calm conditions the only indication that the mother isn't a flattish rock are her three or four exhalations every 15 to 20 minutes — the white columns billowing 5 metres skyward. A tiny brown 'island' fussing around the larger one is of course the calf — sometimes just resting at the tip of mum's snout, or actually lying on top of her head and rolling around in playful movements. Several times a day a calf will suckle, positioned right underneath the mother and nuzzling the mammary slits either side of the genital slit. Evidence of a calf feeding is its tail extended outward from the mother's belly. The calf stays underwater for two to three or even five minutes at a time during feeding.

Even when not feeding, the calves are often right next to the mother, wedged in the crook of her pectoral fin. From this position we have seen the calf pop straight to the surface just like a little cork! Observing the nuances of the tender attention calf and mother pay to each other remains one of the most treasured gifts of our decades of research.

We are also enthralled when the little newborn calves get active and bounce around trying out new behaviours, known as surface active behaviours. These little calves can get very busy and sometimes make much work for the patient, buoyant mothers keeping them safe. We always want to tell them to save their energy for the swim to the Antarctic.

into its life, she was fiercely caring for her newborn. The mating urges of the males could wait.

o

Humpback whales spend the Southern Hemisphere's summer season from December to March feeding in the Antarctic. During the austral autumn–winter, they migrate northwards for the breeding season from April to November, travelling along the coastlines of South America, South Africa and Australia. Off Western Australia, humpback whales migrate northward during April to July, following the edge of the continental shelf in 100 to 200 metres of water. Mating takes place along the length of the coastline, while calving, as we discovered in the early 1990s, is associated with the tropical Kimberley region during the month of August. A non-calving female will mate several times with several males during the course of a winter breeding season, successfully conceiving by the end of the season in October. By December she will have migrated south to the Antarctic to feed on krill for the summer, only leaving the rich feeding grounds once she has adequate blubber reserves to nourish herself and her developing calf for the next 12 months. July through to September is the main calving season off the Western Australian coast. Most whales will aim for the Kimberley, but with more whales in the population now, this trend is becoming less well defined.

With increasing global humpback whale populations, thanks to worldwide protection policies established by the International Whaling Commission in 1963, there are more outliers on the bell curve of birthing-time statistics. As the Western Australian population steadily increases, higher numbers of whales are observed giving birth earlier and later in the season, and in slightly different places along the coast.

When each pregnant female leaves the Antarctic feeding grounds is important in this equation. This timing is set by the ticking clock of her developing foetus and balanced by a conflicting urge to feed and therefore make more blubber (and milk). Her instinct is to migrate north to a warm calving area so that there is less metabolic strain on the newborn calf, but if she doesn't have enough blubber to make milk for the complete return journey, then she and the calf may not make it back to the feeding grounds before her energy reserves are depleted. Mammals living in water, even warm water, must burn enormous amounts of energy to maintain their warm core temperatures. Humpback whale calves are born almost without blubber, and need to consume huge quantities of milk (200 to 300 litres per day) to build the insulating layers of blubber that will eventually sustain life in the freezing polar waters, and it must reach those waters if its mother is to find enough food to survive. The mother does all the hard work initially and the calf need only guzzle her milk, but after 11 months of gestation and a further eight months of nursing, there would be a steep and potentially fatal learning curve for first-time mothers who don't get their timing just right for this 12 000-kilometre round trip between feeds.

The bottom line is that the female spends almost a year producing the calf and then invests most of another year caring for it. She nurses it and protects it from marauding males and killer whales in the tropics, and when it is around four or five months old she will take it to the Antarctic for the summer feeding season. They will head north together in autumn, then they will separate in early winter. So while a female humpback whale invests two years in each calf, in contrast the males invest at most two days with a female while mating.

We marvelled at the unwavering courage of the cow, the sheer pluck of the calf riding her pec fin, and the energetic and overwhelming tenacity of the three males over the course of the one and a quarter hours we observed them. As we moved away, the tussle showed no signs of letting up. This was nature in the raw. It is documented that male lions will kill cubs that are not their own when mating with a female, and I have often wondered if male humpback whales might also practise infanticide. Good thing the females are much bigger, I decided.

During August there are humpback whales everywhere in Camden Sound — playful cow-and-calf pods, quietly courting male–female pairs, rowdy males in combative groups searching for females, more competitive pods of males chasing cows with calves, and even lone singing whales. The first pod off the rank one morning comprised five adult humpback whales: one female followed by four male suitors. Soon the pod increased to seven animals, and after two hours I counted 12 humpback whales all up. Blows billowed continuously as eight, nine or even 10 individuals surfaced at once. Their huge noisy exhalations, like perfectly positioned stage mist, shrouded these dark brown leviathans as they surged, ducked and raced around in the silty turquoise sea. That was 12 times 40 000 kilograms of bone, blubber and muscle — 480 000 kilograms of mammalian drive. Don't for one second underestimate the drive of a sexually interested humpback whale. A whale's agenda during the breeding season entails finding a mate and keeping one, even if only for a short time. At the two-and-a-half-hour mark only two whales remained, the female and a male. It seemed she had made her choice.

There are several theories regarding the mating system of humpback whales. One is called dominance polygyny, where the males, through fighting, develop and maintain a hierarchy for access to females. Another is lek polygyny, where males assemble and display to females through song. A further theory describes male coalitions and cooperation to access females. We imagine we can see examples of each on any given day in Camden Sound!

As the sun set, drenching Camden Sound in glorious tangerine hues, we sought an anchorage for the night; near Wailgwin Island was a perfect spot. We left the final pair together — after all, this was their bedroom, and it was the breeding season.

These gatherings are all about making more whales. Humpback whales are secretive in their mating events. Despite thousands of researchers studying humpback whales across the world, and populations of humpback whales flourishing globally, no-one has witnessed humpback whales mating. This must occur at night, or perhaps it is relatively quick (we observed a possible very hasty coupling of two humpback whales while filming *Birthplace of the Giants* in the Kimberley), and so it remains a mystery. This aspect of humpback whale behaviour continues to intrigue me. We had studied humpback whales for 20 years before witnessing a birth; we might wait another few decades before seeing a mating event. We are patient researchers, in this for the long haul. Among all the other questions filling my mind, I wondered how long the winning male would hold that primary escort position with the female, given there were so many other males around.

I loved experiencing the pandemonium of the throng and observing the antics within the pod. Three males can take the positions of primary, secondary and tertiary escort in relation to the focal female, while yearlings and animals four to five years old, without direct access to the female, often take it upon themselves to swim close to *Whale Song* and make lots of threatening trumpeting sounds. Seeing the bodies of the males covered in old scars and new scratches, some even bleeding, as they jostle at the bow is always exciting. Especially as Curt drives carefully and deftly becomes accepted as part of the pod. He says it's like watching horses or dogs, after a while you get to know every subtle nuance of their behaviour, their moods and mannerisms, and you can read the pod.

By the end of that day we had notes on the datasheet for 11 pods, 35 whales, 32 adults, three calves and 16 photo-IDs. In the middle of all the counting, photographing and note-taking, we had also heard a singer singing love songs to attract females. Other days there are more pods, more whales, more calves and more photo-IDs. The volume of whales matters but the usage of the area as a calving ground is of greatest importance. Areas considered critical habitat, such as calving/breeding grounds and feeding grounds, deserve the highest levels of protection. We knew that this was our purpose.

ONE CRAZY WHALE

It was 7 September 1990 and we were based on Enderby Island, camping in the CALM research station. Our nearest neighbours were the Rothschild's rock wallabies. It was going to be one of those really calm, beautiful days that cetacean biologists live for. Given the good weather forecast for the Dampier Archipelago, I had been up at 5 a.m. getting our gear ready for an early departure. I was glad the extra effort had paid off. The offshore water was very clear, a bright turquoise blue.

In the glassy conditions, Howard Rosenbaum, a visiting Fulbright Scholar, was perched beside me on the boat's pontoon as we peered into the water. A native New Yorker, Howard had joined us at the very beginning of his year-long research project assisting humpback whale scientists across the globe. Beside *Nova*, our inflatable 5.3-metre research boat, two adult humpback whales circled.

One of them appeared to have Curt in its sights. Curt was in the water photographing this rather overly friendly humpback whale. We wanted to know its gender to interpret the unfolding behaviour pattern. For the previous hour, this pod of two humpback whales had circled our boat. The one closest to us, a female, from Curt's observation, was very interested and extremely curious.

There are three easily visible differences between male and female humpback whales. In females, the genital slit is located closer to the anus and is flanked by a pair of 4- to 5-centimetre mammary slits. At the tail end of the genital slit is a grapefruit-sized hemispherical lobe, perhaps used as a visual cue or physical guide for males. Indeed, it was this round lump that Curt had seen clearly as the whale swam in ever-tighter laps around him.

As she swam we distinctly felt her gaze focusing right on us. Whichever eye was presented was actively open as she tilted her head towards us as she made her passes. Sometimes she looked through the water, which made her eyes appear even wider, and at other times she peered at us with her head exposed to the air. For her intriguingly 'crazy' behaviour, I named her Crazy on the datasheet.

With her cetacean 'fish-eye' vision, Crazy was making use of a range of adaptations for her in-water life. Hearing, touch and echolocation are the main senses used by whales, but vision is also employed. The slightly flattened eyeball has an enlarged pupil to allow in maximum light. In addition, the cornea is less curved than that of land mammals and is able to let light in without causing much refraction. A reflective layer within the eye known as the *tapetum lucidum*, also employed by cats in night hunting, maximises light by redirecting it back through the retina so that it is reflected twice. As Crazy ogled us, often with her head oriented sideways, one eye totally out of the water, her pupil was naturally shrinking to protect her eye from the bright rays of sunshine at the surface.

Given the greater number of rods versus cones present in whale eyes, it is believed whales see mostly in black and white rather than colour. We knew Crazy could see us though the water and the air and she was definitely looking. What was she thinking? Why was she so interested?

After the first hour it seemed this whale encounter had moved up a notch. She was getting even more curious. Her movements produced an utterly graceful underwater ballet as she got closer and closer to our inflatable and to Curt. It was very unusual behaviour. She wasn't angry or threatened, just extremely interested. As biologists studying wild animals, we always try to minimise the impact we have on our subjects. But when the animal you are studying turns the tables and studies you, things get funny.

Howard and I couldn't believe what we were seeing. While Curt was photographing one of the four other interested whales that had joined the circus, he duck-dived down just as Crazy came towards the surface. For several moments it seemed as if she was going to surface beneath him, or under the vessel, or perhaps squash Curt against the rubber hull. Was she being careless, conniving or amorous? Curt wasn't about to take any chances and plopped back aboard before she could make him the meat in the sandwich.

'She just came so close!' he said. 'It felt as though she was going to touch me ... She really is crazy!'

Meanwhile, her investigative behaviour continued. Several times, carefully and without touching the drifting boat, she surfaced only 2 or 3 metres away from *Nova*, her large rounded belly facing the deepest part of our boat's hull. Her belly was right beneath us, the ventral pleats, the lines that allow the expansion of the belly from the lower jaw to the umbilical during feeding, were clearly visible. This was insane! On both sides of our boat, she extended her 5-metre long pectoral fins, holding them in the air at a 45-degree angle. What was she doing? Our view from inside the boat was very interesting but I would have loved to have seen her antics from another boat. We were in a very strange position — we had a 40 000-kilogram humpback whale attempting to hug our boat.

One small movement by this wild animal could reduce our trusty vessel to a burst balloon. The acorn barnacles, in clusters the size of a fist all over her body, *de rigueur* for humpback whales, had five razor-sharp calcareous plates that could easily pierce the tubes of our inflatable. More clusters of barnacles on the underside of her chin, along the length of her belly, on the trailing edges of her pectoral fins and on the tips and scalloped edge of her tail flukes were all potentially lethal. Would a friendly whale be our undoing? Is this how our lives would end, the three of us floundering in the offshore waters surrounded by ever-present sharks? As we watched her moving her pectoral fins slowly but extremely carefully in the open air on either side of our boat, I could not believe what we were experiencing. I really did wonder if this was the end, but at the same time I was hypnotised by her antics.

Her endearing yet slightly erratic behaviour intrigued all of us. At first I tried to continue acting the biologist with my clipboard in hand, noting all her behaviours. I counted her rolling and, well, right at the beginning we counted 20 of those, but there were close to 100 just in the first hour! Next, I quickly drew her 'Round' shape and not outstandingly different dorsal fin on the individualised sketches on the pod sheets, along with her accomplice's details, including each animal's behaviour. But the unusualness of this encounter quickly evaporated any possibility of remaining serious scientists. All three of us laughed and giggled in pure delight. It was hard to comprehend what was happening. At any moment Curt was looking into the eye of an adult humpback whale from our port side, while beneath the pontoon on the starboard side of our small boat Howard was getting a great view and then, as I peered back at her at the stern, she took great interest in our outboard motor, which we had shut off some time earlier. We scrambled around the boat following her gaze and movements. She was a very funny whale.

Now with her pectoral fins outstretched so very close to the boat, this whale went even further and leant one pectoral fin against the boat's bow. Without hesitating, Curt reached out to her, placing his thumb and index finger on the leading edge of her huge fin. Gently he squeezed a hello. Crazy reacted by gently moving her fin away about 5 centimetres, then carefully bringing it back another 15 centimetres *closer* to us and then draped her pec on our bow. Our Novurania boat, an Italian inflatable, had a beautiful triangular mahogany bow apron surrounded by two rubber tubes. It was on this wooden piece that the wild humpback whale lay her fin. Given her very measured response, I thought it seemed OK for me to try as well. Our research does not entail touching wild animals, and we conduct our work under strict state and Commonwealth whale research permits because inevitably we are often quite close to cetaceans. But in this case we figured *she* had crossed the line and not read the rules! With an adult humpback whale leaning its pectoral fin on our small rubber boat we were in a position of great vulnerability. If this 15-metre whale decided to react with a full body movement, our boat would easily be capsized. And we would be kaput, end of story.

Nevertheless, I moved quietly and slowly towards the bow and extended my hand as Curt had, wrapping it gently around her pectoral fin. The leading edge of her fin was remarkably thick, about the diameter of an orange. It was beautifully and perfectly curved. The skin was soft to the touch, yet firm and strong all at once. You could feel the muscle beneath the outer skin layers — this was a swimming and underwater flying machine. The leading edge was lumpy, the bumps being the tubercles that provide lift while swimming. From the abundant barnacles, little cirri-covered legs twitched out randomly to capture plankton. I touched between the tubercles. Interestingly these lumps were patterned, and the two along the edge were larger than the others spaced a third of the length apart. Inside this leviathan of bone, baleen and blubber seemed to be a whale with an insatiable curiosity or — at the risk of anthropomorphising — an animal with a very warped sense of humour.

Again Crazy moved her fin away and back even closer.

Howard also held Crazy's fin. She was registering just the pressure of the pads of two fingers ... This was remarkable. We had come here to do serious science, to understand the size of the Western Australian humpback population and to determine their migration paths and critical habitats along the coast, noting depth and water temperature characteristics, but here we were, the first season of what would become over three decades spent along this coast, with a whale draping its fin across our boat and she was responding to our touch! It has never happened again, and maybe never will, but we like to think Crazy reached out to touch the right people that day. (Howard Rosenbaum went on to complete his PhD focusing on humpback whale genetics, and to this day is still protecting whales globally from his base in New York.)

Cetaceans travel in groups that were originally called pods, and this term can refer to one, two, 15 or even more whales swimming together. At the beginning of this encounter Crazy had had the attention of a male accomplice, but during our observations four other males joined the pod, all intently interested in her. Her original companion was able to hurry the others along, a task made easier by Crazy's lack of interest in them. Apparently she was one hot humpback whale — the local boys were thoroughly smitten!

As we made the one-hour bouncy ride back to our mooring in Home Bay on the north-east corner of Enderby Island, I felt extremely grateful. Carrying the two 9-kilogram waterproof Pelican cases filled with camera gear and hydrophone equipment across the tidal flats to our five-star research station, somehow it seemed we had been very lucky yet again. We had returned home, our small inflatable and our lives intact, plus we had had a beautiful encounter with Crazy. This wild whale could have written a very different ending to this story.

Despite wanting to think she was interested in us — and we did get a sense that she was — *really* we were just a distraction. By diverting her attention to our boat Crazy was escaping the advances of at least four more interested males. She was a very clever whale. She was perhaps already impregnated, and had no need nor interest in enduring the advances of more testosterone-charged males. She strategically moved belly up under a boat that kept her genital region out of the sight of these males. It was fortunate for us that the males, despite their fiery interactions with each other, were considerably more intimidated by the boat than this female. We learned a lot that day about humpback whale behaviour and we've since observed this same pattern of females, sometimes with their calves, hiding beside or beneath our boats from extremely interested males — just never with the same level of intensity that Crazy showed.

o

Over dinner that night we explained to Howard that this experience was not typical. We told him he should just fly home right there and then, as such an encounter was incredibly rare and he wouldn't ever get better. He laughed and we all knew it was a great privilege. We had entered rarefied air with Crazy.

That night and throughout the next day, everywhere I looked I could see Crazy's pectoral fin draped across our bow, and once again felt the beautiful firmness of her skin.

A WHALE IN PEBBLES AND MUD

The saying goes 'a pig in mud' but, surprisingly, 'a whale in pebbles and mud' also holds true. Northern resident killer whales, one of four ecotypes in the Pacific Northwest (the others being Southern residents, Bigg's and offshore killer whales), have been documented rubbing their bodies on pebbles off the tree-lined beaches of the British Columbian coastline. One of several known locations is Bere Point Beach on Malcolm Island, situated adjacent to a national park where this unique and culturally important practice can be seen. The family pods come towards the bay vocalising loudly and swimming quickly. Once there, one animal at a time cruises sideways, their lateral body flanks literally rubbing on the small, smoothly rounded, loose-stoned shores of the steep-sloped temperate north-west coast. Some whales just scrape their bellies, maintaining an upright orientation during their rubbing session. When they are done, the pod swims away slowly, the excitement of the event waning. This practice is only done by the northern resident killer whales and has been deemed culturally important. Most certainly a way of sloughing away dead skin, these exfoliation 'treatments' are effective and apparently appealing.

Killer whales are currently recognised as one species with at least eight distinct forms. On the other side of the planet they travel to the warm tropical water along the west coast of Australia, perhaps to reduce yellow-toned diatom coverage they accumulate in the cool Southern Ocean.

In the azure waters of Bermuda two humpback whales were recently recorded in drone footage rubbing on the sandy seafloor. The researcher, Andrew Stevenson, reported that they 'lost' the whale while working with the animal, but the vision through the clear water showed it rubbing against the bottom and thus cleansing its skin and even enjoying it, perhaps just like the killer whales in the north-west of North America.

One day as we bobbed in our research vessel in Exmouth Gulf collecting photo-ID images, Romeo, a male humpback whale in a pair of adults, was courting Juliet, a slightly larger female. They swam close together, rolling around at the surface. We were intrigued and then we noticed the clumps of fine, pale grey clay spread all over Romeo's head and body. Was he packing his numerous fresh wounds, commonplace for male humpback whales, with healing mud? We'd seen this mud-rolling behaviour a year earlier in Camden Sound, also with a scarred and bleeding male humpback whale. Clay is a natural, versatile substance with remarkable skincare properties. There was a time when it was widely used by ancient human cultures for beauty, health and healing but it has fallen by the wayside in modern times. Maybe there are certain types of mud, perhaps from places with freshwater seeps like Camden Sound and Exmouth Gulf, that have special healing qualities? Or maybe the freshwater mud helps to soothe the stinging?

We have so much to learn from the whales. Let's learn as much as we can.

ON THE BELLY OF A WHALE

It was calm weather, typical for August, and here we were in 2013 on board our current vessel, *Whale Song*, on the turquoise, silt-laden water of Camden Sound. The water was glassy and flat as a tack, and there were humpback whales absolutely everywhere. We had deployed a triangular field of three sonobuoys and were monitoring the hardware with the fantastically effective software. With a battery life of eight hours for each sonobuoy and little to no current at that time, there was nought to do but sit back, push the record button and listen to the trills, gurgles and squeals of humpback whale songs swirling through the waters of the sound. The wheelhouse was full of their songs. It was beautiful.

However, curiosity got the better of us. As it was so calm and there were so many whales around, we launched Curt's beautiful hand-built 5-metre fibreglass rowing skiff, *Krill Seeker.* Curt, Simon — our tough, super-fit second mate — and I climbed aboard and we paddled gently away from *Whale Song* to collect photo-IDs from a pod nearby.

Half a mile away we spied a resting whale, its dorsal fin at the surface, the dark body stationary. On my datasheet I ticked the behaviour category 'Resting'. Yes, you could safely say this whale was resting.

Facing the back of the boat to row, Curt adjusted his wing mirror to show the view in front of the skiff. We all knew where the whale was and Curt made a couple of slow even strokes in its general direction. The boat skimmed effortlessly across the water and soon we were halfway there. I clicked away on my camera taking photos of the left dorsal fin and lateral body, what we call a left lateral body photo-ID, from this singleton humpback whale. On the bow Simon filmed with his GoPro camera. All was fine and the whale remained resting at the surface. When we were about 400 metres away, Curt stopped rowing and we drifted.

I looked down to check my photos and make some notes about this snoozing animal. When I looked up again, it was nowhere to be seen. A calm, glassy sea surrounded us.

'Where's the whale?' we all chorused at once.

Then I looked over Curt's left shoulder.

Filling the waterscape around us was an unusual vista. Stretching ahead of the skiff and either side of the boat, and even behind us for at least 10 metres, were the white ventral pleats of the underside of this whale's belly. We looked at each other in shocked silence. The whale had surfaced beneath us — belly up!

The pure white throat pleats were nine to 10 centimetres wide, and their grooves paralleled our boat and extended fore and aft. As the whale settled at the surface with its chin plate exposed and lower jaw just beneath the surface, the skiff gently nudged against its ribs. Against its ribs! We couldn't believe it — we had become beached on the belly of a whale. I looked into the water from the stern of *Krill Seeker* and saw the beautiful white expanse of the animal's underside only millimetres below. I saw the strong tailstock and then the tail flukes. Oh my! I could also see the headlines: '*Researchers drown, flung from boat by whale.*'

Was this the way we would die? Lifting my camera and untangling the strap from my plait, I put it down on the seat beside me. I was certain this whale would lift its flukes and we would be unceremoniously swimming in these crocodile-filled waters... *Sorry, camera, you have been great, but I am sure you are going to get wet, we are all going to get wet.* Now I was talking to my camera.

Almost without breathing we waited for the whale to roll or lift its tail fluke right beneath our light skiff. As the seconds passed we relaxed a little. Would there be a reaction? I picked up my camera and resumed taking photos. The white ventral pleats were like a ploughed field surrounding us. The tail flukes, now several metres from the stern, were only a metre under the water and they gently flexed up and down, creating small swirls on the surface. We moved off half a metre but with the swirling of the whale's gentle movements the skiff was brought back to nudge the exposed ribcage a second time. We had gone aground again momentarily on the belly of a humpback whale.

Surely the whale would react to this? I remembered our encounter with Crazy all those years ago and how she

WHALE SNOT

I distinctly remember one humpback whale that appeared to have a warped sense of humour, or perhaps it brought out mine! On a calm Kimberley day in the Buccaneer Archipelago near Proud Island, we were working with a singleton — a lone sub-adult whale. Sub-adults are teenage whales, basically, animals aged between a yearling and four to five years old. Like our human pre-adults, they often engage in random and erratic behaviour. This whale decided to surface around and around *Sousa*, our 3.5-metre inflatable, and with each blow of its exhalation/ inhalation cycle, it rolled sideways towards us, sending its vapour all over us.

'Don't breathe in the blow!' I cautioned our crew.

Humpback whales, like all cetaceans, and in fact all mammals, possess a vibrant flora of bacteria and, at times, viruses. Directly breathing in the exhalation is not recommended. After several of these close passes around our stationary vessel, this whale gave one particular blow with an accompanying *'Pwff-pwfff-pwff-pwfff!'*, which we noted on the datasheet as 'trumpeting', that was unusually memorable. A great cloud went over us and as we desperately held our breaths, I spied several blobs exiting one of the two blowholes. Somehow one of these globs of mucus landed on my right arm. What? I have been snotted on by a whale! Wow, does this whale have a cold? What manner of infection could it be sharing?

With major advances in molecular research in the last decade, the host and microbial community in, on and surrounding a host organism (collectively termed the 'holobiont') and its functions (its microbiome) have become of great interest. By collecting the vapour and mucus expelled with each exhalation, the health of populations can be assessed from the microbes of an individual cetacean's microbiome. Being snotted on would have been a first-class opportunity to learn about that sub-adult humpback whale's microbiome. In a new collaborative project with Dr Megan Huggett and Dr Michele Thums, we now purposely seek these plumes for analysis. So now we actually *want* to get snotted on.

had responded to our finger-squeeze on her pectoral fin when she had laid it on our bow apron. Luckily she had reacted positively and moved her fin even closer. Why was she curious? What had she sensed? Would we be so lucky with this whale?

For several more minutes this whale lay belly-up with us just floating centimetres above it. Without appearing to even move a muscle, imperceptibly slowly the whale sank lower and lower in the water. It just went down. Several minutes later, in a period of time that felt closer to a lifetime, it surfaced 100 metres away. After three blows it continued to rest, floating, the body from the blowholes to the dorsal fin exposed once again in the warm sunshine.

In sheer relief we all screamed out, 'Wahoo!' We were pinching ourselves, trying to work out if we had drowned or survived. Or both. It had been a rather interesting experience being on the belly of a 35 000-kilogram, 12-metre whale. We were euphoric and amazed at the whale's calm and mellow behaviour.

What had happened? We had been pretty quiet in the skiff. Perhaps the whale hadn't heard us? Though that seemed unlikely. And somehow as it came to the surface with its belly exposed, we were nearby. What were the chances of this? Were we like a blob of floating seaweed to it? We had seen humpback whales actively playing with seaweed. Or had the whale done this deliberately? Keeping in mind Crazy's behaviour, we had a feeling this whale knew where we were and was unfazed by our proximity. However the situation occurred, it was totally unintentional on our part and definitely not recommended. The guidelines, written and enforced by Australian state and Commonwealth governments regarding how the general public approaches marine mammals, stipulate remaining 100 metres away. If your vessel is in neutral and the whale comes closer, this is on the whale's terms and that is fine. Whale-watching vessels all around Australia adhere to these guidelines, hoping that whales will 'mug' them, and quite often they do.

Since 1990 Curt and I have operated our research program in Western Australia under state and federal scientific permits and ethics permits to study cetaceans. The type of work we do often requires close approaches to whales for photo-IDs, biopsy sampling or satellite tag deployment. We take the privilege of working with whales seriously and in any of our projects the welfare of the animals remains paramount.

OUR KIMBERLEY DREAM

Our desire to migrate northwards with the humpback whales to the Kimberley began in the early 1990s as we started to understand their migration patterns. Our initial research work on Enderby Island had detailed the southern migration from August to October that passed within 15 nautical miles of the Dampier Archipelago. But where was the northern migration path? And where was their breeding ground? Clearly it was further north than we had first thought.

To find out, from mid-June to mid-July 1992 we spent four weeks camping on the Montebello Islands, 70 nautical miles west of the Dampier Archipelago, and documented where northbound humpback whales were heading using our 5.3-metre research vessel, *Nova*. The whales, it turned out, were swimming on a bearing of 25 to 30 degrees past the northern tip of the 100-odd islands in this group. This put the whales' northern migration 40 to 50 nautical miles north of the Dampier Archipelago and on track for the remote Kimberley. We had suspected this and had already begun thinking about how we would test this theory.

One afternoon at our three-tent campsite at Goanna Heights on Bluebell Island, the exposed beach provided a perfect canvas to draw a plan of a sailboat some friends in Dampier were building. Before the tide washed away our life-size rendition, we agreed that we could use a vessel of just this size and calibre to conduct whale research in the far-flung Kimberley. But could we build one?

It was John Lally, the director of the Pilbara Camp School in Dampier, who finally convinced us that we *could* build a research boat to follow our research dreams to the Kimberley. He and his wife, Lyn, were building their own 12-metre sailing catamaran, so John suggested we could learn by following their build process, which would conveniently be six or so months ahead of our own schedule, and even more conveniently, located right beside theirs. We were sold! We were about to become boat-builders. The opportunity was too good to pass up.

Over the course of the next two-and-a-half years we beavered away building *WhaleSong*. From the very outset we were addicted. We had a vision and from this we did not waver, despite the difficulties, particularly with funding.

When we were just over a year into the build of *WhaleSong* I learned I was pregnant with Micah. The adjustments were immediate. We had planned a series of whale surveys that year, taking advantage of Woodside's daily helicopter flights transferring personnel to and from the offshore oil-rigs. The very next day I was scheduled to fly to Perth to do another helicopter ditching course, but Curt would have to do this course *and* the helicopter flights instead. Right away our lives were set nicely upside-down!

As my waistline expanded, I often wondered how it would be with a baby on board. I was certain all would be OK, though no doubt nerve-racking. *I mean, think about it, a baby and then a toddler on a sailing boat … in one of the most remote areas on the planet … Just how much more unsafe could it possibly be?*

Despite the prospect of so many new things happening at once, surprisingly, I wasn't too daunted. Perhaps I was in blind denial and perhaps it was all too overwhelming, or exciting, but I sincerely appreciated the wonderful support of our near and far family, my mum and Curt's parents. Also the kindness and encouragement of John, Lyn and the staff of the Pilbara Camp School in Dampier where we built *WhaleSong*.

Each day as we worked among the rocks and spinifex we dreamed of sailing the Western Australian coast looking for whales. Those visions focused our attention as we got covered in glue and epoxy. I imagined crystal-clear water lapping the hulls beneath me while I lay in the trampoline nets. As far as the eye could see interesting white beaches stretched away to starboard, just waiting to be explored, our white sails glorious beneath blue puffy-cloud skies as we sailed …

'*Could you just fibreglass inside that locker, right in the corner there?*'

Ripped from my dreaming, I was back gripping an ice-cream container full of epoxy and a roller. I was Michelangelo with fibreglass! Into that tiny space — and many others — I carefully climbed. My mission was to fibreglass, and thus strengthen, the epoxy cove join between a bulkhead (one of the yacht's walls) and the hull. I liked knowing I was helping to make this boat strong. Yep, with blobs of glue all over our clothes and sticky from dawn till dusk, the idyllic Rottnest Island bays, stunning Shark Bay vistas and remote Kimberley archipelagos could not come soon enough!

Pretty soon I couldn't fit in the small spaces doing those lovely fibreglassing jobs. Phew! That was quite the excuse! 'Sorry, my belly is too big, I am seriously at risk of getting stuck in here forever!'

○

Finally, on 25 July 1995 it was launch day for *WhaleSong*, our custom-built sailing catamaran. At 6 a.m. a truck and a huge crane came to the back of the camp school and our marvellous yacht was loaded for the 10-minute road trip — with a police escort — through Dampier to the marina ramp. As the water lapped the hulls I named her *WhaleSong* and blessed 'all those who sail on her' and whacked the appropriate bubbles on the bow. Our dream had come true!

Three days later, with almost all the important cargo on board — baby Micah, Curt and me, plus skipper Chris Lally, John and Lyn's son, whose carpentry skills had been integral in building this fine vessel — yet still without a functioning toilet, we set sail into a bouncy easterly, bound for Broome. The rousing two days of 30-knot breezes sorted out a few things but mostly gave me about a hundred bruises on my calves! This boating life with a baby would be an adventure, there was no doubt about that.

All the hard work was worth it. All the glue cut out of Curt's hair, all the comments from the Sunday onlookers who repeatedly told us we'd never finish it, the stretched funding and the pressure to meet the first Earthwatch team on 1 August. It was all worth it, especially when we walked along Cable Beach and saw, floating nicely at anchor right before our eyes, our boat built with our own hands.

○

A great sense of euphoria swept over *WhaleSong* each time we travelled into the Kimberley during that first season in 1995. We had a fantastic boat, a new baby — Micah was just seven months old on our maiden voyage — and a skilled and enthusiastic crew training our teams of keen Earthwatch volunteers.

The Earthwatch Institute, formed in 1971, is a Boston-based US company. At that time it provided opportunities for people to volunteer on 50 different research projects all around the world. Earthwatch volunteers assist with all aspects of research, learning hands-on about a multitude of programs in exotic locations. Curt and I ran our Earthwatch programs for six years, and our volunteers were active in recording observations of whales while travelling transects in the Dampier Archipelago, throughout the Kimberley and in Exmouth Gulf, recording the whale sightings in field notebooks and entering the data into our computer database each evening. The data was new and exciting, and for the first time in history reported the distribution and migratory patterns of this population of humpback whales across their breeding and calving grounds. Such a privilege!

During each humpback season, from July to October, we had five Earthwatch teams. Each team comprised five volunteers and stayed with us for 13 days. Over the years we had 150 people assist us with valuable humpback whale research. The Earthwatch offices around the globe handle the bookings and administration for the volunteers, and just over half of the volunteers' fees go to the research project to cover consumables. For our Australian humpbacks research project, this meant the three F's: fuel, film and food. With ready and willing hands, this was an excellent way to make ends meet and conduct field seasons. Along the way we made some great friends.

Curt and I have never been shy about dreaming, and we had big ideas for what we could achieve in our research with our brand-new catamaran, *WhaleSong*. After all, Crazy and her cohorts needed our help, and we'd made a deal and shook on it! Leaving the bubbling metropolis of Broome, which swells from a population of 4000 in summer to 30 000 in the dry season (the Southern Hemisphere's winter months), a world of ancient landscapes filled with whales awaited us.

On that first research season in the Kimberley in 1995 we left Cable Beach and the sheltered anchorage at Gantheaume Point, just outside Broome, and headed north into the deep blue water adjacent to the Dampier Peninsula. Chris Lally, our skipper, had been hired to concentrate on keeping *WhaleSong* safe while Curt focused on the research, operating from a smaller boat we towed behind us or launched over the side. I was on catering and Micah detail as well as whale-spotting duty — all full-time jobs!

In addition, Sam and Fleur, our Earthwatch volunteers, ably helped with logistics on the beautiful Kimberley sea.

Over that first season Micah learned pretty quickly to interpret our stopping, anchoring and shore expedition routine. As soon as the anchor chain began rattling, she would point towards the shore and say 'Beachie?' with a hopeful expression. She loved romping on the Kimberley's beautiful shorelines.

There was no doubt in our minds as to why we were here. We were on a mission to find the northern destination for humpback whales — and find them we did, in between learning about the tides and anchoring in scary places.

Chris and Curt showed a remarkable ability to interpret our nautical charts for suitable anchorages near the coastal mainland and among the rugged offshore island groups. But adding another layer to the unexpected, incredibly, most of our charts, even the most up-to-date ones, featured the alarming word *uncharted* — or still had bathymetry recordings from Matthew Flinders' voyages of 1801–03. This was our first foray into the Kimberley from Broome, and the steep learning curve began at Cape Leveque. Monohulls anchored near the Kooljaman Resort at the tip of the Dampier Peninsula rocked so violently from side to side that Chris and Curt nicknamed it 'Metronome Bay' and perused the chart for a nicer anchorage. Just around the corner was perfect, and we entered it successfully, watching the depth sounder like hawks. It was low tide, and we anchored in 2 metres of water behind the outline of a dune with a spiky-looking top. 'Micah Dune', named for our new baby's spiky red hair, remained a critical staging stop for us for years to come as we waited for the right moment on the tide to tackle the crossing of King Sound.

The Kimberley's 10-metre tides are the second highest in the world after the Bay of Fundy in Canada. With such a huge tidal range, water flows quickly, with currents up to 10 knots between islands and in the bays. This ancient land demands your attention: you need to respect the location and live by the tide tables. We lived for the next three years by the 'Rule of Twelfths', a system John Lally had taught us

BOAT SHOES

It's common for boats to go aground in the Kimberley. Sometimes it is planned ... and sometimes not. We built customised 'shoes' onto the bottom of the hulls of our catamaran so that we could deliberately go aground twice a day with the huge tides, if we wished. This enabled us to efficiently refuel and resupply in Broome. When the tide was out we could drive our 4WD ute onto the sand flats at Town Beach right next to where *WhaleSong* sat safely high and dry. We could literally pump unleaded fuel from a 44-gallon drum on the back of the ute into the yacht's tanks and easily lift boxes of groceries on board, rather than having to anchor several miles offshore and row a small dinghy in and out with precious goods.

This type of grounding was, of course, well planned. All our other movements through the Kimberley were designed to ensure we didn't bump the bottom at all.

for mathematically working out the tides for each hour of the day. 'That's another $5 trade secret,' John liked to say — we must owe him millions!

As we headed north we experienced the curving white pearlescent sandy beaches bordered by the ancient bright red pindan soil and red rocky outcrops surrounded by clumps of tropical pandanus palms beneath the bright blue Kimberley sky. There is nothing subtle about the Kimberley. The colours are intensely strong — they hit you right in the face with a wonderful assault of colour that combines with the beautiful textures of the rock formations and vegetation. *I just love it!*

o

Details of how humpback whales used the waters of the Kimberley were unknown in 1995. We did, however, have some extremely useful sightings from the Coastwatch crew to guide us. Through a fortuitous meeting with Kaye Grubb at a slide presentation at Perth Modern School, one of the Coastwatch pilots, Charlie Grubb, kindly offered to transcribe their whale sightings for us. The federal government's Coastwatch aircraft made frequent flights over north and north-western Australia looking for illegal fishing and drug smuggling, and in the course of this they noted anything and everything they observed — including humpback whales.

These hand-plotted Coastwatch sightings, mostly showing compass bearings and estimated distances to or from islands (GPS was still a novelty in those days), served as a pilot study for places to visit during those first three

Kimberley seasons. Over the first two years, we took each team of Earthwatch volunteers to the five concentrations of whales noted in the Coastwatch sightings hand-plotted on our extended chart. In order, sailing north from Broome, these five places were Quondong, the Lacepede Islands, the Buccaneer Archipelago, Frost Shoal and Camden Sound.

Our second team's survey, for example, went in a circuit from Broome to the Lacepede Channel, then Pender Bay, Cape Leveque, Byron Island, Parakeet Channel, Hall Point, Koolan Island, Cockell Reefs, Camden Sound, Augustus Island, Sampson Inlet, Heywood Island, Lucas Island, Champagny Island, Darcy Island, Wilson Point, Hall Point, Collier Bay, Raft Point, Steep Island, Koolan Island, Cockatoo Island, Yampi Sound, Byron Island, Cape Leveque, and finally – 600-odd nautical miles and 13 days later – back to Broome! Exhausting and exhilarating!

After a wonderful 1995 season full of delights – the islands *really* were that beautiful and there really *were* so many humpback whales doing their thing among the gorgeous archipelagos – I continued to be amazed by our *WhaleSong.*

'Curt, look at that yacht. It's just amazing. Where did we steal her from?'

After two years, 1995 and 1996, we had documented the position of every whale we'd sighted, and were able to confirm the Coastwatch sightings were accurate. Not that we didn't believe them, but since we couldn't know their flight paths (being a federal secret), we could never be certain that these were the only locations of whale congregations. With the determination of explorers like Cook,

Dampier and Flinders, we were beginning to make progress in our quest to put the whale delivery room on the map for conservation management.

Left A field biologist's kit in 1990: Curt and me with hats, sunglasses, camera, compass, binoculars and trusty notebooks. Nowadays research tools have gone hi-tech but good old-fashioned observation is still important. *Photo: Douglas Elford.*

Below I took this photo as the sun began to set while filming the Nat Geo Wild documentary *Birthplace of the Giants* — it was a stunning early evening. We followed this pod of whales until nightfall with me listening from the bow and radioing to Curt the direction of their blows until it got too hairy among the scattered islands and reefs in the darkness. *Photo: Micheline Jenner.*

Above Earthwatch volunteers were integral in our exploration of the Kimberley, observing, sighting and plotting whale movements. Half of each volunteer's payment to the Boston-based group went towards the three F's — food, fuel and film — that kept *WhaleSong* afloat. *Photo: Micheline Jenner.*

Below It took 23 months for Curt and me to build our catamaran, *WhaleSong*. Our two girls were raised on the boat and it took us on many adventures, including our discovery of the remote humpback calving grounds in the Kimberley. *Photo: Vanessa Sturrock.*

Our daughter Tas, then 11 years old, took this photo of a 22-metre pygmy blue whale from the 12-metre mast of our current *Whale Song*. On deck we are collecting photo-ID and positioning to deploy a satellite tag. *Photo: Tasmin Jenner.*

With dolphins on the bow, Skipper is never far away! His barking does not deter Indo-Pacific bottlenose dolphins, like these, that will bowride for 30 minutes or even an hour or two at a time. *Photo: Micheline Jenner.*

Above For 20 minutes this sleeping humpback whale remained stationary with the tips of its tail flukes exposed in a secluded cove at the northern end of Koolan Island in the Kimberley. *Photo: Micheline Jenner.*

Below This little humpback whale calf swims through the blue, emitting a tiny stream of bubbles from its paired blowhole. *Photo: Curt Jenner.*

Above This humpback whale breached over 100 times in a row! Breaching is a surface active behaviour and can be a means of asserting dominance in a pod or scaring away predators — or even ridding their bodies of hitchhiking suckerfish. Besides all that, breaching also looks like a lot of fun. *Photo: Curt Jenner.*

Right A Jenner family shot on the ice-class vessel *Whale Song*. From left, Tas (12), me, Micah (16) and Curt. Below us, peeping through his favourite fairlead, is Skipper, ship's dog extraordinaire. *Photo: Frances Andrijich.*

Above In the surreal Antarctic light, this humpback whale calf breached over and over again. Usually humpback whales refrain from this kind of surface active behaviour on the Antarctic feeding grounds, but this little calf had some learning to do!
Photo: Micheline Jenner.

Below Antarctica: on a trip following the coastline explored by Douglas Mawson 102 years before, Tas, me and Curt hold the Explorers Club Flag 69 and honour the explorers of the past on *Whale Song*.
Photo: Inday Ford.

PROTECTING THE KIMBERLEY

The Kimberley is a special place. This tropical bioregion is part of the huge Indo-Pacific realm, where the waters of the Western Pacific flow into the north-eastern Indian Ocean via the Indo-Pacific through-flow, which then mixes with the Indian Ocean. The extremely varied coastline — rocky ria shores with inundated gulfs, a mountainous hinterland and long stretches of beaches — borders a rich ecosystem of diverse and species-rich coral reefs and mangrove habitats that are home to dugongs, saltwater crocodiles and sea turtles. This is all in addition to the offshore marine habitats with their dolphins and whales.

The Kimberley's vast inland area of some 423 000 square kilometres is spectacularly beautiful and so rugged it remains a true wilderness region. On the offshore islands, huge rock faces of red and gold tones mix with patches of beige, and spinifex fields overtop piles of white rocks. Large flat, pink river stones characterise Langii at Freshwater Creek on the mainland, and round pink stones identify the Macleay Islands to the north-west, suggesting it was perhaps once the mouth of Freshwater Creek. The mainland is different from the islands, and each island, although with similarities of shape and geology to its neighbours, also has its own distinct flavours.

The Kimberley's interior extends from Port Hedland in Western Australia to the Northern Territory border, and is one of the world's largest wilderness areas. This is an ancient land — the natural processes forming its geology are estimated to be 1.8 billion years old. Dinosaur footprints on the beach at Gantheaume Point, Broome, and at James Price Point, date back 130 million years. Fossils consisting of remnants of stromatolites and blue green algae discovered at Marble Bar are 3500 million years old.

The idea of reserving marine areas for environmental protection and resource management began in the 1960s. The premise was that the ecosystems to be preserved were to be representative of regional habitats and biodiversity. In the 1970s the Western Australian Environmental Protection Authority promoted a policy of 'Conservation through Reserves', and in 1975 the National Park Authority began planning the Ningaloo Marine Park. Since those early days, the Western Australian coastline has had 16 Class A *Conservation and Land Management Act* marine reserves gazetted, in addition to seven *Fish Resources and Management Act* fish habitat protection areas, the Scott Reef Nature Reserve, and a marine protected area (MPA) around Rottnest Island. One of two further MPAs recently added to the suite of reserves in Western Australia is Lalang-garram/Camden Sound Marine Park.

The rationale for this large MPA was not its representativeness of habitats and biodiversity in the Kimberley. It was because of our discovery that Camden Sound is a principal breeding area for the Southern Hemisphere's largest population of humpback whales.

Curt and I are thrilled that the Western Australian state government has protected the 'bedroom' of this humpback whale population. It was our ultimate goal in beginning this study. But it required strong persuasion and a bit of luck. When we first confirmed the Coastwatch humpback whale sightings during surveys on *WhaleSong*, we made the appropriate scientific and permit reports but kept the area a secret. And we kept it secret for 15 years, fearing that this secluded location would be overrun by interested and well-meaning people and ruined for the whales.

Our Centre for Whale Research data, stretching back to the mid-1990s, was the foundation for the government's understanding of the region as a critical habitat for the humpback whales. Still, there was a need for impassioned campaigning for the cause by Broome-based organisations such as Save the Kimberley, Kimberley Whales and Kimberely Media (Richard Costin and Annabelle Sands), and the Wilderness Society. Professor Jessica Meeuwig, inaugural director of the University of Western Australia's Centre for Marine Futures, was integral in negotiating several protected areas within a larger boundary.

In 2012 Lalang-garram/Camden Sound Marine Park was gazetted. The main species the park is designated to protect is the humpback whale, however, as is typical of whale conservation, many other species in the area also benefit from its protected status. While this is a win for the whales, at 7000 square kilometres it only represents the portion of the calving ground that falls within state waters. An area of equal, if not slightly larger, size adjacent to the marine park that is just as important to the survival of the humpback whale population lies in waters under federal jurisdiction, and for various reasons has yet to be protected. Don't worry Crazy, we knew it was going to be a long haul.

Education is the key. In 2014 we deployed a satellite tag on a pregnant whale just west of Camden Sound while filming a National Geographic documentary, *Birthplace of the Giants*. This whale, with her newborn calf in tow, provided evidence of the full range of habitat by swimming an almost perfect track around the perimeter of what actually should be the boundary of the marine park. Unfortunately political will comes and goes with these sorts of things — a shame, since the whales just need a safe and quiet place to give birth to their calves, nothing more.

AN ANCIENT LINEAGE

In the ancient landscape of the Kimberley, 'living dinosaurs' — crocodiles — inhabit the murky coastal and offshore waters. And whales also have an ancient lineage.

Popular belief holds that terrestrial mammals, either mesonychid condylarths (wolf-like, hoofed animals) or artiodactyls (even-toed, hoofed mammals that include ruminants such as cattle, camels and hippos) were the ancestors of the cetaceans, the whales, dolphins and porpoises.

The theories state that when terrestrial mammals ranged from the land into the warm shallow water, gaining a taste for food within the sea, these triggered functional changes for full-time life in the sea. From the first radiation, cetaceans are represented in the fossil record 34 million to 55 million years ago in the Eocene era within the warm, shallow tropical waters of the Tethys Sea. The second radiation led to the odontocetes (toothed whales) and the mysticetes (baleen whales) around 23 million to 34 million years ago in the Oligocene era. The third phase, about 5 million to 23 million years ago in the Miocene era, led to the modern whales, particularly the delphinids (animals in the Family Delphinidae, the marine dolphins) and balaenopterids (animals in the Family Balaenopteridae, the rorquals, which includes blue whales and humpback whales).

THE ORPHAN

It can be incredibly moving to watch the gentle caresses between cows and calves. We have seen a calf delicately stroke its mum with its floppy pectoral fin, then nuzzle her with the tip of its head. The tenderness and carefulness they show towards each other can only be described as affectionate. Patient mums tolerate all manner of bumps and clumsy love, even when their calves plop right on top of their heads! These tactile behaviours appear to be the basis of a process of bonding and caring which extends into the first year of the calf's life.

On this particular day the turquoise Indian Ocean near Legendre Island in the Dampier Archipelago was beautiful, calm and clear. We were bobbing in *Nova*, our 5.3-metre workboat, and observing a cow and calf humpback whale pair. Patiently we waited for the adult whale to surface. Some 25 metres from us the small calf circled fretfully. Its swirling swim pattern was about twice its body length in diameter, and it seemed as though it was desperately searching. Surprisingly, given the distance, we could hear this little calf making squeaks and trills, but these were not the usual social sounds produced by cows and calves, which can include whispers. These were panicked calls, and their persistence and loudness were alarming. This calf was distressed. Apparently we weren't the only ones waiting for mum.

A colleague from the Western Australian Museum, Pat Baker, a world-class maritime archaeologist and underwater photographer, had joined us for this week in October 1992. It was the third field season of our humpback research project based on Enderby Island. On calm days I always woke the boys early, making lots of loud 'Let's get going …' sounds from inside the shack while preparing breakfast and making our cut lunches. As usual, I aimed to get us on the water by at the latest 7 a.m. in order to use every single moment of daylight to work with the whales while maintaining our momentum and stamina across the physically demanding five-month field season.

On this particular day, we had already worked with a few pods of migrating humpbacks, collecting the usual left and right lateral body photographs and opportunistically presented tail flukes. Most were heading south, as it was mid-October and the southern migration was proceeding as usual. The whales were doing their thing and there had been nothing out of the ordinary so far.

Life on Enderby Island was very physical. We lugged heavy equipment every which way. Like ants we trekked over sand dunes — and across 200-metre tide-flats when the tide was out — at the beginning and end of each research day. If the tide was in, we luxuriously rowed a

3-metre dinghy in and out of the bay to a mooring, then made the sand dune stagger across the island. Not to mention that the middle of the day was filled with up to 90 nautical miles of bouncing around in our small research boat. Accordingly, our in-field winter calorie requirement exceeded our in-town summer needs. Because we were always rather hungry, Curt's default way to bring a whale to the surface always involved food. If we seemed to be waiting a particularly long time, perhaps for a singer or a sleeper, Curt would threaten to start consuming his beloved sticky PBHs — peanut butter and honey sandwiches.

'OK, it's time to get my hands totally gooey and have a PBH. This ought to bring the whale up. Can you please pass me the sandwich box?'

On this occasion Curt finished his sandwich, washed his hands, had a slurp of water and was ready to go. But we were still waiting. Where was this whale?

We had approached the cow and calf pair and slowly paralleled them. Unusually, we had seen the calf fall a little further and further behind. Curt had swung the boat around and we slowly approached the calf. Surely the mum would be there. Perhaps the calf had been suckling and soon the cow would surface.

Well, Curt started looking around for more things to eat! All the waiting was making him peckish. An apple was procured. Then a few handfuls of GORP (good old raisins and peanuts), our boat version of trail mix. Soon the minutes turned to tens of minutes and I couldn't believe it when we saw the blow of an identifiable adult whale almost two nautical miles off to the west ... Did that

HUMPBACK ALTRUISM

Recent research has documented the unusual behaviour of humpback whales protecting their own species — in the Kimberley we had seen how mothers protected their calves. But humpbacks have also been recorded protecting other marine mammals from killer whales. One theory for the humpbacks' long migrations from polar feeding regions to tropical breeding zones is to avoid killer whales while calving. Certainly encounters with killer whales do still occur, but over 100 cases have been documented of humpback whales actively protecting other animals from the killers' advances. Humpbacks can use their long pectoral fins not only to fly through the water but also to bat animals away. These appendages can also be a haven, as in the case of the little calf tucked into its mother's pec fin. There is a fantastic and, for humans, endearing photograph of a seal tucked inside the pectoral fin of a humpback whale as it fends off the hungry advances of a killer whale. This behaviour sparks the question: are humpback whales altruistic?

Dr Bob Pitman, a cetacean researcher with extensive shipboard survey experience, gathered observations of this protective behaviour and concluded that the answer was yes: humpbacks were indeed altruistic. They care for their fellow humpies as well as others in the marine community. Wow, as if we needed any more reasons to want to protect humpback whales!

mother just ditch her calf? Why? Then an awful realisation hit: maybe this wasn't the mother of that calf. So where was its mother? What had just happened?

As we formulated these questions the calf continued circling frantically and producing increasingly distressed sounds. We watched as it headed over towards a pod of two adults that were migrating north. Due to the configuration of the coastline and the location of the islands of the Dampier Archipelago, a whale on the northern migration is in fact travelling eastward, and conversely, a whale on its southern migration is actually swimming in a westward direction.

This pod or group of two adult humpbacks were males. We could tell by the large amount of scarring on their bodies. Battle scars are gained by males competing for females on the breeding grounds during the warm, tropical winter months. Male humpback whales migrate to the calving grounds to mate as often as possible with available females. A union with a female may last two hours or two days — there is no commitment by the male to establish a family unit, and plenty of scars are a just reward.

The calf confidently swam alongside this pair of man-about-town types. We stayed half a mile back, but close enough to observe what unfolded. The calf swam boldly and quickly with these adults in the first surfacing sequence, but it soon became apparent that this little poppet would cramp their style. Seriously, why would they want a calf with them? With each subsequent surfacing, the pair appeared half a mile away from the calf, either to the left or the right. As the calf bolted to catch them, they dived again, and so it continued. These males were going against the migratory flow, heading north to maximise their chances of mating with as many females as possible. A calf in tow was not a good look. The males were zigzagging back and forth to ditch it.

By this stage tears were streaming down my cheeks. What could we do? Could we entice the calf back to Home Bay on Enderby Island? Could we somehow net the bay? Could we try to feed it? As I proposed each idea, Curt sensibly explained why it wouldn't work. Of course I was being unrealistic, but I was desperate to do something.

The calf swam fractiously in random directions looking in vain for the northbound adults. Sadly, they had successfully given it the slip and were merrily travelling at a good pace, between five and eight knots, back towards the Kimberley. Barely over six metres long, a calf of this season, the little whale bobbed to breathe at the surface. It was a tiny grey body on the wide, wide sea. It was on its own and tired out from its two unsuccessful attempts to follow the adults. It looked very little and very lonely.

'Look, Curt! There's another pod! It's two adults and they're heading south.'

I was ever hopeful, and I got the feeling this little tacker was hopeful too. It swam over towards the pair that were travelling slowly on an intercepting course. Just like that, the calf popped up alongside and swam right beside them, their pace much more suitable for it. Could these whales save this little whale? Curt and I presumed they were a male/female pair as we made observations of their behaviour and physical features. They were comfortable together, possibly even a courting couple. One was slightly larger and less scarred and so was likely to be the female, and the other, more beat-up looking, was the male. With some relief, we left this third pod of the afternoon thinking perhaps there was a glimmer of hope for this calf.

What would be the chance that this female could feed the calf? In theory, only a very slim one. From well-documented studies sperm whale calves are cared for and nursed by their mothers and non-mothers in alloparental care, and dolphins also engage in care of others' calves. But these accounts are of sperm whales and dolphins, which are odontocetes. Family ties are greater and totally different from those of baleen whales such as humpback whales. Would compassion cross traditional classification lines? Humans nurse non-biological children as so-called wet-nurses in short-term care and non-mothers have been able to lactate babies in long-term adoptive situations when the unique psychophysical trigger provides the let-down response. I wondered long and hard about this. If this female humpback did look after that calf, which was highly unlikely — and highly unlikely that we could have documented it, since evening and nightfall were about to set in — this certainly would have been a very rare situation. I held fast to a tiny thread of hope.

As we skipped over the waves into a tangerine sunset over the darkly silhouetted red rock islands of the Dampier Archipelago, I knew in my heart of hearts that the outlook for that calf was grim. Humpback calves need to feed several times daily from their mum's rich, 40–50-per-cent-fat milk, and they need approximately 200–300 litres per day for most of their first year of life. Having somehow lost its birth mother, this was not going to happen out of the blue for our orphan. We were watching nature in action. In reality, the calf may not have lasted the night. What *had* happened to its mother?

CLOSE PASS

'Five times five is twenty-five, six times five is thirty...'

Clapping in time with our words, Micah and I were doing our daily times-table routine. Over our racket we heard Tas call, 'Look, she's coming!'

We were all in the main saloon of *WhaleSong*. As seven-year-old Micah and I had clapped away at our times tables, near the door, with a clearer view outside, three-year-old Tas had looked up from her puzzle. Curt emerged from the office just as Tas announced that a whale was getting even closer.

It was a huge female with a young calf, and the pair were heading slowly but steadily towards *WhaleSong*. We'd anchored a little further offshore than usual in Lighthouse Bay as it was exceptionally still and the lack of wind meant the bush flies swarmed around the yacht at Bundegi, the mooring area 10 kilometres south of North West Cape or Vlamingh Head. The mother, healthy and spectacularly rotund, swam directly towards our anchored vessel. They would have spent a week or two resting in Exmouth Gulf, a relatively shallow north-facing bay, and were now continuing their southern migration, winding down the Western Australia coast. She wasn't in a huge hurry. Her two- to three-week-old grey-toned calf bobbed beside her, gently moving in her slipstream.

Curt took Tas onto the aft deck as the cow surfaced and exhaled and inhaled with a huge *whoosh*. A smaller puff of air also escaped from the smaller lungs of the calf, just a metre from the hull. Almost in slow motion the mother swam parallel to *WhaleSong*. Carrying Tas in his arms, Curt walked the length of the hull as mother and child swam beside it. Both the cow and calf looked up at Curt and Tas, and Curt held eye contact with the mother as he paced. Time seemed to stand still even though the whole event only occupied 60 seconds of our day. Despite the briefness of the encounter, Curt felt a shared trust and vulnerability as he gazed into the huge mother whale's eye and felt a renewed strength of purpose to protect these amazing animals. As he held his own child, he again vowed to do whatever it took to protect hers.

As this pair moved away from *WhaleSong* to continue southward, they joined an increasing cetacean stream. Dotted across the horizon outside Ningaloo Reef, the 240-kilometre part-fringing, part-barrier reef extending from Bundegi in the north to Amherst Point to the south, white whale blows billowed skyward.

MOURNING THEIR DEAD

Recent extensive research has revealed that cetaceans mourn their dead. Detailed observations from Dr Cynthia Moss in Africa over 40 years ago indicated elephants behaved remarkably in the presence of the bones of family members. Their behaviour implied they knew the bones and their significance, as with 'respect' and 'affection' they gently stroked them. Cynthia's conclusion was that these animals performed ritual-like 'ceremonies' not dissimilar to the funerals held for loved ones by humans.

Seven cetacean species are reported to mourn their dead with rituals of holding, tossing and carrying deceased pod members, particularly calves. Indo-Pacific bottlenose dolphins, one of the species listed in a recent paper, regularly visit the Fremantle Fishing Boat Harbour. One afternoon while working on deck, Curt noticed a mother bottlenose dolphin. The researchers at Curtin University and Murdoch University had named her Tupac. She was desperately tossing her newborn calf along the axis of its body, almost as if to miraculously wake it. Her first calf, Gizmo, had had a fishing line wrapped around its dorsal fin that cut the fin at the base as it grew and the water flow pulled the tissue loose. Sadly this calf died later that year, in November 2015. What was the problem with this next calf? For at least six hours we watched this mother repeatedly, almost violently, flinging her calf in the air, batting it with her flukes and then swimming beneath it and lifting it up, as if in agonised frustration. She also pushed it around on her nose. Under the orange nightlights of the jetty, she swam 25 to 30 metres away from the calf and then surged towards it at high speed, trying in vain to lift it. It was heart-rending to see, a mother so unwilling to believe her calf was dead and so desperate to revive it. Her dedication was hard to forget.

SUNSET THRASH

We were less than two days out from Fremantle, sailing south after completing a visual and acoustic observation survey from Darwin. Now it was late afternoon and my tummy was rumbling. I could smell all the delicious aromas coming from the galley where Resty Adenir (*Whale Song*'s chief engineer and kitchen wizz) was cooking, and I was looking forward to one of his signature Filipino dishes, chicken and pork adobo served with mountains of steamed rice and a fresh garden salad. When we're at sea doing whale research, I fix lunch for the crew and Resty prepares our evening meal. Sunset was still an hour away, and as I studied the pinkening sky I recalled all the wonderful cetacean sightings we had made in the last 10 days as we travelled down the coast. I quietly thanked the world for the chance to be out here. Despite the tigers growling in my tummy, I continued to scan the sea for the usual signs of blows and bodies.

A dark shape moving two miles offshore captured my attention at 5.30 p.m.

'Curt, I've got something. It's on the starboard beam, two miles.'

Immediately Curt turned *Whale Song* and as we closed the distance we made out a juvenile, perhaps a yearling humpback whale, twisting and turning at the surface with very floppy movements of both pectoral fins and the tail flukes. As we got closer I recognised the distinctive smell of whale blubber and noted a large slick of whale oil surrounding the desperate humpback whale. This lone animal was being chomped, chewed and harassed by several sharks, with many cobia and half a dozen sucker fish circling below it. In the fading light I adjusted the settings on my camera, amazed that I was still able to collect images. Thirteen hundred photos later, I was certain we would have some clues as to the demise of this poor whale. The overall body condition was very good; the whale's girth was rounded, it was not skinny, and it seemed healthy with all body parts present and functioning. However, alarmingly, there seemed to be many white scars near its dorsal fin, as if from some type of entanglement — a chain? A rope? Or even iceberg scratches? Or were they just very unusual pigmentation patterns?

The floppiness of the tail flukes seemed the key to the mystery. Had an entanglement of some kind damaged the whale? Had it tried to migrate regardless, following its internal program? Had something happened that day to affect its tail flukes so drastically, leaving it unable to swim properly? So many questions. How long would this little whale last as it thrashed and rolled at the surface?

It was devastating for Carrie, our research associate, and me to witness this animal's desperation. The whale was clearly frustrated at not being able to move properly and, sadly, its fate was sealed. It wouldn't last the night in its present company.

After an hour of watching this unique situation, we pulled away into the darkness, hoping that the end would come mercifully soon. This whale's mother would likely never know what had happened to her offspring: once it's a yearling, a humpback whale calf separates from its mother. Such is the cycle of life.

The next day we saw a dwarf minke whale and heard humpback whales inshore and pygmy blue whales offshore. Despite these pleasant distractions, my mind constantly mulled over the distressed whale we had encountered the day before. Helpless to assist, we were concerned to understand what had happened.

After trawling through several hundred of my high-definition images, Curt came up with a possible scenario. This yearling whale had been well cared for by its mum, as indicated by the healthy body condition and the intact pectoral fins and tail flukes. So lack of food wasn't the issue. Curt concluded that it was blind and possibly also deaf, perhaps from birth. None of the photos showed the whales eyes open, despite our close proximity. Normally a whale would 'eye-widen' at the close approach of something the size of a small ship, and the sound of our motors didn't seem to change its behaviour either. The scars and scrapes all over its body were testament to either being especially 'clumsy' and banging into mum or icebergs, collecting multiple barnacle wounds and scars, or being blind. The ocean is not a place that forgives weakness.

The beginning of the northern migration is the time for mother and her yearling to separate, and their strong bond is broken. The mother may initiate the timing of this as her milk supplies naturally dwindle. She may wait for the calf to take its usual snooze at some point, then slip quietly away. This is the way of humpback whales. When did this yearling separate from its mother? If it was indeed blind, it would be unable to navigate far without following its mother, so the separation may have taken place that day. The mother may not have even known her calf was blind, since their relationship is mostly tactile. The calf could feed simply by touch supported by the mother knowing what to do. There was little chance this yearling would have lasted the night — a reminder of how difficult life in the ocean can be.

WHALE FALL FAUNA

At some point, all living creatures die. Even the great leviathans of the sea. Observations of dead cetaceans first began in 1854 off the Cape of Good Hope, South Africa, when a centimetre-sized mussel was first seen on the blubber of a dead floating whale. Now a vibrant study of whale falls has revealed 400 species of animals living in and around sunken whale carcasses. Some 30 specialised species are only found in these environments.

Since 1992, a team of dedicated researchers have had the unenviable task of towing dead beached whale carcasses offshore for sinking and monitoring. Three stages have been identified:

the mobile scavenger stage, the enrichment opportunistic stage, and lastly the sulfophilic stage.

In the mobile scavenger stage, hagfish actively tunnel through the meat and sleeper sharks bite off chunks of blubber. Together they take blubber, muscle and the internal organs at a rate of 40 to 60 kilograms a day. This stage can last up to two years, depending on the size of the whale. The enrichment opportunistic stage, which also lasts for up to two years, is dominated by polychaetes (bristle worms) and crustaceans. Finally, the sulfophilic stage is characterised by specialised bacteria anaerobically breaking down lipids contained within the bones. These bacteria use dissolved sulphate (SO_4) as their source of oxygen and release hydrogen sulphide (H_2S) as waste. A 40 000-kilogram whale carcass may contain 2000 to 3000 kilograms of lipids, and their decomposition is a slow process. For a large whale, the sulfophilic stage can last up to 50 years, perhaps even a century. The research team suggests that at any given time there could be 690 000 skeletons of the nine largest whale species rotting around the world and enriching the ocean environment.

Dead whales are an essential part of our ecosystem. Yet another reason why live whales are integral to our life on earth.

SINGERS

'Curt, there's a whale singing underneath us!'

It was bedtime, and I was cleaning my teeth on deck when suddenly the unmistakable moans and trills of a humpback whale permeated the boards beneath my feet.

I excitedly whispered to Curt and any crew in the saloon who were still awake. The song was louder in the hulls, so our assistant researcher Katie and I lay down on the wooden floor and listened to the undulating and repetitive rhythm of a male humpback's mating song. The wooden hulls of our catamaran acted like a guitar; we had a 13-metre wooden instrument transferring beautiful whale sounds from the water to our living space. Our little floating home was resonating. The male singer was after any female whale in the sound. It was simply beautiful. I could picture him in the rather shallow, muddy-bottomed water of Camden Sound. He would be lying virtually motionless, probably within 100 metres of us, his pectoral fins outstretched and slightly rotating to maintain his desired position in the middle of the water column. His body would be on a 45-degree angle to the water surface, his head down and his tail towards the sky. He was singing a love song. I was smitten. As I listened to that beautiful song I decided there and then that if I were a whale, he would win my heart. Well, for a short time, anyway.

We had heard plenty of singers over the years while we were working in the Dampier Archipelago, and we even loved to stick our heads in the water or slip over the side to feel their intriguing songs reverberating through our bodies, but somehow this one seemed special. Standing on the aft deck, I gazed all around. On an inky sky a trillion stars sparkled, the remoteness of the location far from light pollution and a new moon gave the stars their best chance. As my eyelids grew heavier and heavier after a long day, my soul brimmed with the events unfolding below. Purely by being in the right place at the right time, we were listening in to our whales' breeding antics. We often encountered singing males and witnessed another adult joining them. Would this happen tonight? Would it happen right near *WhaleSong*? There was a whisker of voyeurism here — this was their bedroom, after all …

As I drifted off to sleep, I hoped his melodic self-promotion had been effective. I wished for a gentle coupling beneath our hulls and bridge-deck, just as I had seen in Maui. There, in the bright blue sea, a pair of Hawaiian humpback whales appeared to be linked belly to belly. Even now, 30 years down the track, I still have the image of them distinctly in my mind's eye — the strip of a white tail and a black tail pressed close together. The underside of the flukes showed on the humpback whale positioned deeper underwater, and were slightly offset by about five centimetres from the flukes of the other animal, which presented the topside of its tail and body towards the surface. In this motionless coupling they drifted through the water.

o

We especially enjoy working with singers and recording their plaintive mating songs. Whenever we saw a lone individual, there was a good chance it would be a singer, particularly if the whale's downtimes were extended (i.e. more than 10 minutes) as he perfected his love tunes. Humpback whales are renowned for producing the most complex songs, not just sounds, in the animal kingdom. Males sing while underwater in tones ranging from 40 or 80 to 4000 hertz in complex and diverse songs that comprise phrases, sub-phrases and units (the smallest piece) in songs that last from 10 to 40 or so minutes. The unmistakable trills and deep moans of humpback whale songs vary within and across seasons, as Katy and Roger Payne discovered in the 1970s. As well, uniquely, the humpback whales' songs are representative of individual populations in terms of the rhythms of phrases and sub-phrases.

To record a humpback whale's song with our hydrophone (an underwater microphone), we motor our inflatable, *Sousa*, slowly to what we call the fluke-up footprint, the place where the singer has just dived down, turn off the motor and lower the hydrophone over the side into the water. We have to make sure there is no movement of the cord against the hull that could contaminate the recording.

We first captured whale songs with cassette tapes and tape recorders. Later we used digital (DAT) recording systems. As we sit in the boat, we listen through headphones to the sounds of the sea.

Over the course of three seasons in Camden Sound we noticed that the length of songs at the beginning of the season were 11 to 14 minutes, while at the middle and towards the end of the season, the songs extended to 20 to 25 minutes. This was by virtue of the increased maturity of the singers. The younger males, with less mating and life experience, produced shorter songs. The older, battle-scarred mature males sang for much longer. Do females prefer longer songs? It seems likely that males in the prime of life have the best lung power and volume. It would be clear to a discerning female which male was strong and therefore a good mate. But what was the function of the complex and changing songs? Song structure must also be important — as it is for many birds. Is it a test of the male's brains as well as his brawn? Humpback whales are socially complex beings and more successfully recovering from the pressures of whaling than any other species of whale. Are they more intelligent than other species?

One day when we were out recording, I asked the Earthwatch volunteer scribing the data notes, 'How long has that been?'

It was the end of August, and another glorious day in whale research paradise. Curt and I and two volunteers were in our runabout *Sousa*, bobbing around in Camden Sound.

'That's 21 minutes so far, they went down at 10.19.'

LOOKING LIKE YOU: MIMICRY

There are many cases of mimicry in the animal kingdom: a king snake copying a coral snake, a hoverfly trying to be a bee or even jumping spiders acting like ants. The copycat lives in a similar habitat and, with some adjustments to its colouration and behaviour to make it look like something more dangerous than itself, becomes protected by the association.

In cetaceans, a possible case of mimicry exists with two small deep-canyon sperm whales copying their shark predators. In a separate family classification from sperm whales, the *Kogiidae* has two species of *Kogia*: the dwarf sperm whale (*Kogia sima*) and the pygmy sperm whale (*Kogia breviceps*), which are often difficult to tell apart at sea. Sometimes solitary or in pods of up to six individuals, these whales grow up to 2.5 metres and have the unusual appearance of a shark, including a shark-like tapered head with underslung jaw, sharp teeth and, curiously, a light-coloured false gill slit. Our Centre for Whale Research (CWR) team has seen three of these dwarf sperm whales, which were distinguished by their larger, more erect (shark-like) dorsal fins. What is even more interesting about these little whales is that when threatened by predators, they use the ink from the squid they prey upon in a burst of their own faeces in order to create a dark red cloud that allows them to escape.

The turbid turquoise water of Camden Sound was as flat as a tabletop. Nary a zephyr ruffled the datasheet. Around us various pods went about their daily business — that is, the business of finding a friend, keeping a friend, mating, and then, for the female, caring for a calf. Male humpback whales do not form a long-term bond with humpback cows.

A pod of possibly five whales breached constantly in the direction of Augustus Island and towards Wilson Point a cow/calf pair rested like two black islands, one a fifth the length of the other. Near Bumpus Island and skirting Rice Rocks, a pair of adults travelled quickly, perhaps a couple of males coming to investigate the floating mum and neonate.

Sweat beaded on my forehead and trickled down my back. When was this whale going to surface? In the bright still air we were feeling every one of the 37 degrees Celsius of warm Kimberley sunshine.

'OK, Pavarotti …' Actually, this singer was more like Tom Jones. 'Now it's time to come up!' He went through his repertoire of three to four low tones and three to five of the higher, characteristic humpback whale shrieks, which he repeated over and over again. The squeaky door sound indicated an imminent surfacing, and here it was: the whale was coming up. Now it was time for us to get going. We started the motor and headed over to the surfacing whale to get a left dorsal photo-ID and thus complete the three photographs needed to identify this whale. This singer had indeed sung a fine song — in the end it was 48 minutes 13 seconds long!

Collecting humpback whale songs from the Dampier Archipelago during the first five years of our humpback whale research, we noticed that the Western Australian song was clear and simple in composition. Among a myriad of calls, one was very similar to the motorbike frog: a frog that made a motorbike sound. These endearing amphibians inhabit Perth backyards, serenading each other on warm summer evenings with their distinctive rev-like 'ra-ra ra-ra'. In the blue of the Indian Ocean, surprisingly, we listened to a whale sounding like a frog sounding like a motorbike! Was this life imitating art or art imitating life? Other whales' songs could sound as though the whale had severe gas, with lots of high-pitched 'toot-toot-tooting', while others are very guttural grunts and almost expressive moans with warbling trills and squeals thrown in.

One season, we provided the tapes of our Western Australian songs to colleagues Dr Doug Cato and Associate Professor Mike Noad on the east coast, who wanted to compare them with their recordings. The results were earth-shattering. In 1996, two whales on the east coast recordings were singing the simpler Western Australian song from the previous season; in 1997, half of the east coast whales were singing the west coast song; by 1998 all the east coasters were belting out the Western Australian top 40! A paper written with our colleagues made the renowned scientific journal *Nature*. The evidence of such a rapid change in the song was deemed a 'cultural revolution' rather than a 'cultural evolution'. Mike's theory was that male whales on the eastern coastline appeared to quickly adopt the new song with the express hope of winning over females. So it wasn't all about lung power and singing the longest song. Clearly novelty played a role in how the females selected a mate. This implies that male humpback whales have the intelligence to change their songs in the face of competition, and that female humpback whales choose mates that show an adaptive, intelligent streak. Behaviourists across the world were excited by the apparent speed of these changes and we fielded interesting emails for months from all sorts of biologists and even psychologists.

SATELLITE TAGS

We now know that Camden Sound is of great importance to humpbacks, but what about other areas to the north? Is York Sound the next Camden Sound? In 2009 we spent two weeks in the Kimberley with our colleague Dr Mike Double from the Australian Antarctic Division applying satellite tags to female humpback whales in three areas in order to learn how they used the region. The most sensitive gender and age class of a humpback whale population is the breeding females, particularly given that cows provide a year of parental care to their calf of the season.

Mike, Curt and I worked from our small inflatable, with Curt expertly driving and Mike shooting the satellite tags. I took photos of the individuals and the tag attachments. At the stern of the boat, where I was positioned, it could get a little busy taking photos and taking notes. As we approached a group, we immediately assessed the behaviour within it. Were the whales suckling, resting or travelling? We avoided any feeding or resting animals, but those on the move we paralleled. Humpback whales are

BLADERUNNER: A STORY OF SURVIVAL

During our Antarctic voyage in January 2014 we came upon a pod of two humpback whales. As I took left lateral body photos of the larger adult, I could see at least 15 deep propeller scars extending the whole side of the whale ... How had this animal survived? We saw only two surfacings as the pod disappeared into the fog. I thought it would be reasonable to believe that this whale would be weary of vessels, having been injured so badly. Even the edges of the tail flukes had not escaped — three slices made the right tip of the fluke wobble freely.

When we posted the left lateral image of this humpback whale on our CWR Facebook page a few months later, I was not prepared for the response, 'That's Bladerunner!'

Pretty soon we had some of the story. Sydneysiders had got to know Bladerunner in June 2001 when this humpback whale with a vivid white 'zipper-like' ship-strike scar right across its body spent two days in Sydney Harbour after suffering injuries from a boat propeller in Broken Bay.

In 2008 when Cat Balou Cruises sighted Bladerunner off Eden, the body tissue of the whale was starting to protrude between the white strike marks. The next sighting was in July 2013, as the whale headed north alone, and then in October 2013 Bladerunner was seen heading south off Twofold Bay (Eden) with a calf.

This calf, now three to four months old, was with Bladerunner when we saw her among the fog in the Antarctic. Wow, what a survival story!

highly mobile; they are expressive and have a wide range of aerial behaviours. Cows protecting their young can exhibit ferocious tail-slapping and head-slapping towards nibbling sharks, an example of their stroppy surface active behaviours.

Adult humpback whales often do a fluke-flick after the collection of a biopsy sample or the deployment of a sat tag, which we believe would be like a mosquito bite to them — annoying but quickly forgotten. When we are in close proximity to apply the tag, I have learned a few things. Firstly, to take a whole lot of photos of the whale; secondly, to immediately turn around to protect my camera from the ensuing drenching splash; and thirdly, to wear quick-drying clothes. Dripping from head to toe is nicely cooling in the mid-30-degrees tropical climate! It certainly adds spice to the deployment!

Several functioning tags revealed a number of interesting details of the humpbacks' southern migration, including the average distance to the coastline being less than 25 kilometres, high variability of resting/movement among the pairs — two animals went 1200 kilomteres into the Indian Ocean — and one female showed her journey of over 7000 kilometres to the ice edge.

O

'Wobbles is up — this is looking like a good surfacing ...'

A few minutes later the third satellite tag of Pod 3 (the third pod of the day) was deployed on a humpback whale migrating northward near Quondong, north of Broome. This adult was named because of its unusually wobbly dorsal fin, which moved back and forth like the comb of a rooster as it surfaced. We had also noted that this whale set the tone for the animals within the pod. As the first to surface we immediately recognised Wobbles, followed by Dot, Spotty, Chop and Mr Issues. By telling individuals apart within the group we could note which animal had been photo-ID'd or tagged and who was up next.

Over the course of that day in late July 2008, we encountered 43 humpback whales in 15 pods (30 adults, 10 subadults and three calves). Wobbles joined Chop, Spotty, Sub and Chunk wearing satellite hardware, all of them helping us better understand how this population was using the Kimberley.

Replete after a great day with whales, and satisfied that we'd achieved our objectives regarding humpback whale migration, we decided to use the last tag of that season's six for a deployment on a northbound pygmy blue whale near Rowley Shoals, a day's journey west of Broome on the edge of the continental shelf. Blue whales, and their slow recovery from whaling, were increasingly becoming the focus of our interests and we wanted to carry the momentum of our work with humpback whale conservation into a study of blue whales — the largest creatures on the planet. At 10 p.m. we lifted the anchor of *WhaleSong II* and steamed west. From close to seven o'clock the next morning, we were surprised when we began sighting humpback whales, 21 pods, in fact, of 33 animals. All adults, the majority of these whales were travelling north-east. We were halfway between Broome and the Rowley Shoals, so their migration direction was taking them to the northern Kimberley archipelagos, the Buccaneer and Bonaparte and, presumably, Camden Sound.

At around three in the afternoon we picked up a mooring outside Mermaid Reef, the northernmost of the three pristine oceanic atolls of the Rowley Shoals. Mermaid has no permanent land and is thus protected federally, Clerke and Imperieuse Reefs have small islands so come under Western Australian state jurisdiction. The view from our observation tower as we approached was simply unreal. From 20 nautical miles away we had seen the lagoon — in fact we had seen strangely green clouds, the reflection of the green lagoon on the underside of the clouds.

And so, 'Please take me to the green clouds' was my new wish. Using clouds as clues of location is not new; the whalers in the Antarctic looked for dark blue on the bottom of the clouds to indicate clear and open, non-iced-over water. The traditional Indonesian fishermen have also used the green reflections on the clouds for several hundred years when they sail southward without a compass for six days to fish in the shallow lagoons of Scott Reef.

A day later, on our return journey to Broome, having not sighted any blue whales, we saw even more humpback whales — 61 animals in 39 pods, again all adults — at this same halfway mark. A new secret was unfolding when we least expected it. Our entire career has been about finding and interpreting the elegant patterns of these master navigators. Being offshore in dedicated surveys for decades has yielded much data but, in the end, just exploring offshore in random locations, such as this trip out to Rowley Shoals provided information we hadn't even known to look for. The split in the migration path at these latitudes became important years later when oil and gas companies wanted to conduct seismic surveys in the area at the time when pregnant females would be approaching the calving grounds. Getting to the calving grounds is as important as being protected in the calving grounds, and we were happy to have added another piece to the migratory puzzle for future management. The next day we saw more whales and, being closer to the coast, their direction of travel was mixed but there were still no calves — it was too early in the season. The learning never stops.

Standing on the platform of our observation tower looking towards the tropical lagoon, there was time to be introspective and tears streamed down my cheeks. The water of the lagoon glowed mauve and pink. The shallow lagoon was *so* clear, and the porites coral bommies encrusting the seafloor below coloured the water. Not a speck of dust or dirt sullied this lagoon. It was unbelievably crystal clear and so beautiful it took my breath away. I thought of my mum and wished we were together to experience this beautiful sight. Phil Bouchet, our research assistant, also thought of his mother living in the faraway Swiss Alps. As we swam in the lagoon we enjoyed it vicariously for our dear mothers.

How privileged were we to experience these incredible sights while contributing to cetacean research?

Humpback whales had been our focus for 10 years. We had shown that the population was recovering steadily post-whaling, we had mapped most of their migratory paths and resting areas along the Western Australian coast, and most importantly we had discovered and mapped the boundaries of their calving grounds and influenced the government to create a sanctuary there for them. But our

MIGALOO: AUSTRALIA'S WHITE WHALE

A white whale as beautiful as an iceberg, Migaloo was first photographed passing Byron Bay on the New South Wales coast on 28 June 1991. Observers initially thought this humpback whale was an albino, but currently Migaloo is considered to be 'hypo-pigmented'.

This totally white whale, part of the eastern Australian humpback whale population, has captured the imagination of the world with its beauty and uniqueness. Sightings along the coast spark Migaloomania, with the excitement spilling over into news stories, Facebook posts and tweets as tourists and whale-watchers scramble to see him.

The White Whale Research Centre provides interesting information on Migaloo at migaloo.com.au/.

Due to his uniqueness, Migaloo garners a police escort in New South Wales waters, and along the entire east coast special distance restrictions apply to boats hoping for a closer look, with hefty fines for those who disobey.

Travel safely, Migaloo! Imagine finding Migaloo in the Antarctic, a white whale in a white environment ... That would be insane!

energies were peaking and we weren't content to sit back, especially when other species of whales were not faring as well. Our concern turned towards the magnificent blue whales, protected from whaling for even longer than humpback whales, yet it appeared their numbers were hardly recovering at all. What was going wrong?

BLUE WHALES

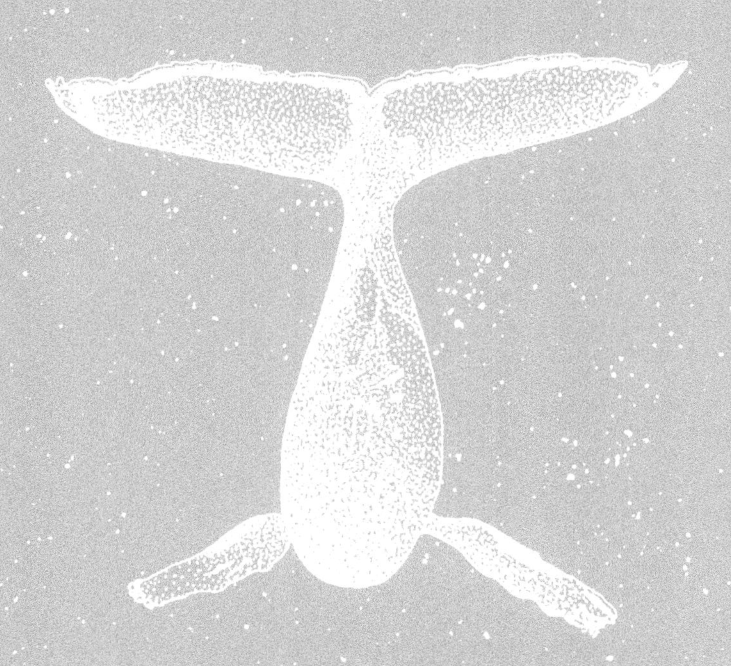

WITHIN THE BLUE

Wearing a thin summer wetsuit, I slipped over the side of our small Zodiac into the cool, deep blue water of the Perth Canyon. We were some 74 kilometres west of the port of Fremantle in Western Australia; the capital, Perth, was just another 20 kilometres inland along the Swan River. Convergent golden streams of light pierced the ever-increasing indigo of the 1300-metre deep water. I was transfixed by the deep and the blueness, but I had things to do: a pygmy blue whale had just dived. I was swimming behind it, through its footprint — the fleeting round swirling pattern left at the surface from the flukes when it dives. I could see a light dusting of tiny white pieces like snowflakes falling slowly through the water column. Searching the edge of this deep underwater world through my mask and breathing evenly through my snorkel, I wielded a small aquarium net to collect some of these little white squares.

Mammals routinely shed dead skin cells known as slough skin. Pygmy blue whales, like all cetaceans, replace their outer skin cells frequently, making collection a benign and potentially useful tool. Genetic material can be contained in these sloughed skin cells and we were collecting these flakes from identified animals each time they dived. At the time there was ongoing debate among scientists about the amount of useful genetic material slough skin contained due to the lower quantity of live DNA in these dead external cells. Nowadays with the rapid advances in genetic research and techniques for analysing 'touch DNA', this is actually quite a viable alternative to taking biopsy samples for particular species. When a blue whale dives, it makes a final exhalation/inhalation and rounds out to expose its dorsal fin, flexing its gargantuan muscular body. This procedure releases clouds of slough skin, making them good candidates for this technique.

This blue whale had a lot of loose dead skin, which I supposed would feed untold millions of zooplankton and small fish. Following about 30 metres behind I could see the whale's pale blue body diving efficiently down into the blue without appearing to move a muscle. I quickly caught a netful of the roughly thumbnail-sized pieces of white skin and was just about to swim back to the boat with my precious items when I came across an area of golf-ball-sized, bright orange blobs. Now this was something new to investigate. When I touched the edge of a blob with my fingertip, the mucous-like outer skin immediately broke, creating a diffuse, tangerine cloud in the water. Raising my head to alert Curt, who was in the Zodiac nearby, I inadvertently put my head through another soft conglomeration of the floating material.

'Curt, what's this orange stuff in the water?'

The orange stuff had now settled on my head, in my hair and large globs slid slowly across my forehead and down the centre of my mask, staining my field of view. The smell, even through the mask, was horrendous! We both quickly realised it was whale poo. And it was everywhere. It was all around me and all over my head. It was like being coated in soft salty cowpats.

'It's poo, Curt! It stinks!' I yelled, trying not to get any in my mouth.

It was hilarious and we both instantly knew we had reached another defining moment in our careers. This whale, one of the least common and least understood on the planet, had defecated … because it was feeding. This told us that the Perth Canyon was a unique place, a feeding ground for pygmy blue whales — we were on to something.

As we followed alongside the whale I took in all the nuances of its appearance. As the pale grey rostrum at the tip of the head broke the surface, the expanse of the flat upper jaw became visible. Enormous! Next a tall blow billowed to nine metres above the surface with an almighty '*Whoosh!*' Exchanging close to 95 per cent of their lung capacity, cetaceans are efficient breathers and the sound carried the force of an explosion. After the blowholes opened with the exhalation and closed again post-inhalation, the pale grey central ridge of the backbone arched in a shallow curve, almost parallel with the water surface. On and on this rolled, blotches and spots peppering the expanse. Finally, three-quarters of the way down its back, a small pointed dorsal fin, quite understated, broke the surface.

WHALE POO

An interesting theory has come to the fore in recent years regarding the fundamental role of whales within the ocean ecosystem. Whale poo, rich in nitrogen and iron, fertilises the phytoplankton (algae), which blooms and is fed upon by zooplankton (krill and all manner of fish larvae) and small fish (herring and capelin), which whales forage upon. The migrations of whales move these nutrients, including iron, throughout the water column and across ocean basins, mixing them effectively. By reducing whale stocks such as humpback, fin and blue whales during the decades of whaling in the previous two centuries, the nutrient levels of the oceans were also depleted. The world's oceans need the complex mixing of nutrients that large populations of whales provide. The fish stocks we depend on would be healthier and more abundant if the bottom of the food chain was healthier. Species of whales that feed in polar regions, like humpback whales, also mostly defecate in these areas, and as a result they self-fertilise their own food stocks, the krill, and globally their populations are doing better than any other species of whale. It could take much longer for other whale species that feed in more dispersed and temperate areas to recover and rescue themselves in this way, and our commercial fish stocks are likely to remain low until then.

'Flukes are coming!' I cautioned Curt, who was standing in the bow with the camera poised and ready. From the dorsal fin to the tail flukes is the caudal peduncle. Arching the dorsal, almost forming a triangle above the sea, the thick, muscular caudal peduncle was lifted at least 3 metres above the surface. The crowning glory was the 7-metre-wide tail flukes with their relatively smooth trailing edges and pale grey undersides. Water had flowed across the topside of the flukes and then poured off the edges in a momentary 7-metre waterfall. My eyes brimmed with that view.

That night when we returned to Longreach Bay at Rottnest Island (12 nautical miles from Fremantle), we horrified and delighted the girls with our poo story. Needless to say, I doubly lathered up all over before jumping into the bay for my usual saltwater and soap 'shower'. I was certain my shipmates would not want to share the dreadful smell.

In our previous 10 years of studying humpback whales on their northern and southern migrations to and from the tropics in Western Australia, we hadn't seen humpback whale faeces because the whales were focused on breeding rather than feeding. These blue whales were different — they were feeding. This was a eureka moment!

The lives of baleen whales are seasonal and usually highly structured, with summer for feeding and winter for breeding. During winter there are two alternative strategies for baleen whales, one that results in fat whales and one that results in streamlined whales. Fat whales, such as humpbacks, grey whales and right whales, store blubber to live on for part of the year. They usually congregate in well-defined locations during the winter, where calves are born in warm water and males take advantage of the clusters of females to battle for available mates. The adults don't feed in these places. Calm, warm water with minimal disturbances is essential to quickly transfer energy via milk to young blubber-less calves before they start their long migration towards the adult whales' feeding grounds near the poles. Living off their fat reserves from the summer feeding season, the adults fast for the entire winter breeding season. By season's end they can become almost anorexic, having lost between 3600 and 4500 kilograms during their winter migration. The reason why humpback whales undertake the largest migrations known in the animal kingdom still eludes scientists, but their system works — they are the most successful species of large whale on the planet at the moment in terms of population numbers, despite humanity having brought them to the point of near-extinction twice in the past century.

By contrast, the more streamlined rorquals, such as blue whales, fin whales, sei whales, Bryde's whales and minke whales, do not store huge amounts of blubber nor do they clump together in great herds, but instead roam the oceans far and wide, searching for food year-round. These whales take advantage of various seasonal food supplies that alternately boom and bust, and their fine, streamlined bodies are efficiently designed for travelling on the long searches that are sometimes necessary to find these intermittent ocean-cafés. One downside to this strategy is that mates are less abundant if everyone doesn't go to the same place at the same time, so areas that have enough food to attract and maintain seasonal concentrations of these types of whales are few and far between, at least in today's oceans.

When Curt and I and our colleagues began documenting blue whales in the Perth Canyon in 2000, little was known about why they frequented this area. Our research group was not even sure which species of blue whale we were encountering, and we were certainly unaware that these whales were feeding. But from the second I got covered in poo, we knew what these whales were up to. This was a krill café, and pygmy blue whales were stopping to feed here. Amazingly, only a few of these places were known to exist on our planet, let alone in Australia, and our team had found one right here in our own backyard!

On our next trip to Fremantle for supplies we sent a canister of poo to the faecal analysis lab at the Australian Antarctic Division in Hobart, where samples from all manner of marine animals are expertly analysed by scientists. The combination of dead krill scooped at the surface, live krill found swimming in the water column near the surface and the just-pooped faecal samples solved the puzzle. The mouth-parts of the live and dead krill specimens were compared with those found in the faecal material and the conclusion was made that these pygmy blue whales were indeed feeding in the deep waters of the Perth Canyon. The krill in the samples, specifically the species *Euphausia recurva*, is a 10-millimetre long crustacean which was being consumed, along with as many as 27 other krill species that have now been identified in the Perth Canyon, at a rate of up to 2 tonnes daily by each hungry pygmy blue whale.

o

Five years prior to my life-changing encounter with the orange poo, our friend and colleague, John Bannister (JB), large-whale expert and then director of the Western Australian Museum, had encouraged the International Whaling Commission (IWC) to investigate the population status of blue whales south of Australia, where historically the Russian whalers had taken thousands. The IWC ship steamed out of Fremantle bound for the Roaring Forties with John aboard as the chief scientist. To everyone's amazement (except John's I think!), large pale grey-toned whales were sighted by the scientists only half a day out of port. They reported whales as 'like-blue' and 'blue whale' with 'up to two per day recorded in December 1995 in a small area some 25 nautical miles off Rottnest Island, in the area of the Perth Canyon, Western Australia'. JB regaled us with his stories of these rare and intriguing animals just over the horizon from our base in Fremantle. He'd found none anywhere else in his survey of the Roaring Forties whaling grounds, and our interest was instantly captured.

The next summer we watched the weather patterns carefully for calm wind and sea conditions in between bouts of the notorious 'Fremantle Doctor' sea breeze to make a trip to the canyon. This afternoon sea breeze from the south-west predominates the daily weather patterns of south-west Western Australia in summer. It is so named for its regular daily doctor's visit to cool the city of Perth.

At dawn on a warm and glassy February day with easterlies pushing us briskly offshore, we sailed towards the canyon aboard our sailing catamaran *WhaleSong*. As we travelled westward the easterly wind eased and we rose and fell gently over the 1-metre sou'-westerly swell. Bow-riding beneath the nets of our catamaran were common dolphins with their striking white and cream crossover markings on their flanks, delighting Micah, then two.

We looked for further signs of whales and dolphins from as high up as possible on the cabin roof.

'There's a blow!' I blurted, pointing.

Sure enough, there it was, and then another tall, straight columnar blow reached skyward, allowing Curt and my brother, Ross, to see what I had seen. Half an hour later I saw another blow, this time in a different direction, and the unique low-angled blow in the light wind conditions well and truly indicated the presence of a sperm whale. Both sightings were between 3 and 4 nautical miles away from us, but the blows and the species they indicated were clear as a bell. We had found at least one blue whale and one sperm whale, two uniquely deep-water species, and we were in the mix! With night approaching we had to turn around and head for home without close approaches to either species, but we were satisfied the Perth Canyon was a unique place just on our doorstep. Logistically it wasn't going to be easy to study here, as it was a long way offshore in a windy part of the coast, but it was going to be *good*!

In that first season, we went out either in *Mega* (our 6-metre Zodiac) or *WhaleSong* to investigate the blue whales in the Perth Canyon. There were so many questions. The IWC sightings had directed our attention to this area, but now we needed to start from scratch and pose the right questions.

What sort of blue whales were they? Pygmy blue whales or true blue whales? Nowadays true blue whales in the Southern Hemisphere are called Antarctic blue whales. Which species or suite of species frequented this area? Where had these whales come from? How long were they here? Where would they go next? How many of them were there? It was fascinating for our team as every investigation yielded new data. Our Western Australian consortium of scientists, all close friends for many years, wrote a proposal which succeeded in obtaining funding from Environment Australia (then the federal wildlife department) and the Department of Defence. A new multifaceted project was born. With JB, Associate Professor Rob McCauley, Dr Chandra Salgado Kent and Dr Susan Rennie of Curtin University, Chris Burton of Western Whale Research, plus Curt and me from our Centre for Whale Research, we planned six years of work, starting with broad-scale aerial surveys and fine-scale boat surveys to work out how large the area was that these blue whales used, and what they were doing there. The aerial surveys would also attempt to estimate the numbers of whales that used the area each year. Simultaneously, a variety of techniques would be employed for the fine-scale boat surveys, again focusing on population numbers, using both photo-ID and biopsy sampling, as well as general behavioural observations of the whales and size estimates to establish their age class. We knew we had our work cut out for us — but this would be *amazing*.

During the whaling era, blue whales were highly prized and targeted most frequently, since the yield per unit of effort was greatest. One blue whale could yield at least twice as much oil as a humpback whale (on average 70 to 80 barrels versus 35 to 40 barrels) and killing one whale was easier, and ultimately cheaper, than killing two. Given a choice, whalers in the Southern Ocean deliberately sought out these mightiest of animals as their quarry. A black-and-white photograph taken beside the hull of a whaling ship records the largest animal ever caught — quite possibly the largest animal that has ever lived. This female Antarctic blue whale was killed in February 1928, and was recorded as measuring 30.5 metres and weighing 160 tons.

Over the summer of 1993–1994 I had taken part in an IWC shipboard survey in the Southern Ocean near the soft ice-edge of the Antarctic. With two other scientists, Paul Ensor and Luis Pastene, I spent 13-hour days conducting visual surveys and noting all cetacean sightings in this extraordinary landscape of swan-shaped icebergs, wandering albatross, the rose hues of the midnight sun and the tell-tale splashes of racing minke whales. During this cruise I was fortunate enough to have had seven encounters with eight Antarctic blue whales.

One of these experiences I will never forget. As snow softly fell, our Antarctic survey ship travelled steadily behind a pair of enormous whales. I stood on the bow, shielding my video camera with a waterproof cover, almost unable to believe my eyes. The pair kept a good pace, moving effortlessly ahead of the ship at around 15 knots. The notes for this encounter list the adult female blue whale as measuring 24 metres and her calf 15 metres. It was clearly a calf of that season, presumably born in the previous July or August. Endearingly, the two swam very close, side by side. The mother's sheer size, her graceful movements and her steady resolve to guard her calf will stay with me forever.

The promise of blue whales near the Western Australian coast, and the chance to participate in a project documenting blue whales near a metropolitan city, held huge scientific and personal interest. When we began this research, there were only two long-term blue whale projects being conducted globally, one off the California coast by Cascadia Research, and the other by the Mingan Island Cetacean Research group, based at the St Lawrence River, Canada.

One of our first experiences on this blue whale project in the Perth Canyon remains vivid in my mind. We were observing every nuance of these animals to gain basic behavioural and biological information, so each time we saw a blow we changed course towards it in the Zodiac and carefully observed the size and shape of the whale's head and body in an effort to increase our understanding of the physical features and differences between the two subspecies, the Antarctic blue whales and the pygmy blue whales.

We had sent biopsy samples for laboratory analysis by Dr Bob Brownell in California, who confirmed that these were pygmy blue whales. However, we were also keen to determine in the field, via detailed visual observations, whether each animal we encountered was indeed a pygmy blue whale. Perhaps the Perth Canyon was important for both subspecies? However, after repeated encounters, year after year, we confirmed that the Perth Canyon was primarily a feeding area for pygmy blue whales. Visually, pygmy blue whales are, as their name suggests, smaller than their larger cousins by some 5 to 6 metres. However, their heads are relatively larger, giving them a 'tadpole' shape in contrast to the 'torpedo' shape of the larger subspecies.

On this particular research day, huge splashes halfway to the horizon attracted our attention. As we approached we realised this was a relatively large pygmy blue whale. As it surfaced, it was at least three times longer than our 6-metre Zodiac, so we estimated its size to be 22 metres. It was surging through the surface of the water, creating a massive white wake either side of its head as it raced through the ocean. I raced along next to it while Curt stood in the bow taking photographs. If I strayed to the left or right in any dramatic fashion, Curt would simply fly right out of the boat.

'OK, Curt, hold on, we are going 24.5 knots, now 25.6, I'm trying to stay straight!' I called over the roar of the motor and the slap of the waves.

My heart was pounding. It was crazy — this whale was moving so quickly, we couldn't believe it. With each surge at the surface we had a clear view of its profile. Its large, dark-toned head was surrounded by huge, explosive waves of white foam. Moving in line behind the whale, we could also see it pumping its strong, short tailstock. I never imagined I'd witness a 'porpoising' blue whale, let alone be driving right next to one, but here I was. We're not sure where this whale was going in such a hurry or, alternatively, what it was running from, but it was an awesome display from such a magnificent animal at full stretch. We couldn't keep up with it in the end and collapsed in the bottom of our Zodiac with aching thighs and exhausted lungs.

DON'T WE KNOW YOU?

'Look, this one has a remora on the fluke!'

Our research assistants Vanessa Sturrock and Kate Fitzgerald were chatting in *WhaleSong*'s office during our 2004 blue whale field season. Vanessa was commenting on the tail flukes of a pygmy blue whale in a black-and-white image that had been transferred from a colour digital file into the grey scale with 256 shades, within which the human eye can recognise the smallest detail.

During our fine-scale research into pygmy blue whales in the Perth Canyon, we collected photographic images of the left and right dorsal fin and tail flukes (if presented when diving) of each individual to count the number of animals feeding in the canyon. Not all blue whales fluke-up when they dive, and not every surfacing reveals the dorsal fin, so photo-ID is not as straightforward as it is for humpback whales. However, it didn't take us long to understand that the technique simply involved a bit more patience. We needed to wait through up to 15 shallow-breath recovery dives for the whales to begin a deep, sounding dive before lining up for the perfect ID photos. Each deep sounding dive event, and thus photo, could take up to 20 minutes, followed by a five-minute interval at the surface for the whale to recharge its oxygen supplies. Humpback whales go through this cycle in half the time or less. We worked out our system from closely watching the blue whales' swimming technique when the all-important last breath before a long dive was about to occur. Following the whales in the clear waters of the Perth Canyon, sometimes from the trough of a huge swell while the whale was on the crest, was easier on this final surfacing. We realised it was all about physics and momentum and that as they glided along, their enormous pectoral fins would flare outwards from their flanks in order to steer their elegant, streamlined bodies towards the surface. For the deeper sounding dives, the pec fin flare was exaggerated, so the whale's back and tailstock, and sometimes even the flukes, were pushed up higher out of the water. Seconds later their momentarily airborne mass would propel them almost straight down into the deep for another feeding session. There was real rhythm and efficiency of motion, like a rolling wheel, describing these enormous swimming machines (just as 'whale' comes from the the old-English *whael*). Photographing them is always a pleasure.

'You know what, we had one a few days ago with a remora in the same spot as well. I'm going to look for it.'

While I worked through the books doing the accounts, I heard more about the blue whale photo-IDs that were being entered into our customised computer matching and archival system. Vanessa Sturrock (who by then was joining us for the third time while putting her geology career on hold), searched the database for that particular shape of fluke, as well as the animal's unique dorsal fin. Sure enough, there it was: it had been a week between sightings, but we had the same photo seven days apart, complete with what appeared to be the same dangling remora. This pygmy blue whale was still feeding in the Perth Canyon and it still had a remora attached in exactly the same place!

Another unique marking we observed on another whale was a huge shark bite in the caudal peduncle. As this pygmy blue whale made the last blow of a surfacing sequence and dived for the deep, the back arched, the dorsal fin was raised and the caudal peduncle was lifted in a fluke-up dive — to reveal a huge half-watermelon-shaped scoop was missing from the dorsal ridge of the peduncle, just in front of the whale's tail flukes. Looking at the photos that evening back on *WhaleSong*, Curt scaled the wound pixel by pixel against the approximately 5- to 6-metre length of the whale that was visible between the dorsal fin and the notch in the centre of the tail flukes. The bite (which was what it appeared to be) measured at least 750 millimetres across, possibly as much as a metre. Curt pulled a tape measure out of the drawer and we giggled nervously as we tried to imagine a mouth that could take a single bite that size. Swimming in the canyon took on a new perspective, and how the whale survived this injury we don't know. But two years later, in 2002, we saw it again, so we realised these whales are tough, despite all sorts of threats.

BLUES FROM A BLUE SKY – LOGISITICS RULE

For six years every summer we did an aerial survey every 10 days, looking for blue whales in the Perth Canyon. I soon learned there were quite a few layers involved in getting airborne each flight. Firstly, the twin-engine Partenavia plane that was suitable for offshore work had to be available to hire from the Royal Aero Club at Jandakot. Next, Gail, our pilot, needed to be free from her full-time shifts as a paramedic. Then there had to be no military exercises underway in the area. Chris, our aerial team leader, had to be available, *and* the weather needed to cooperate —no more than 18 knots of wind for optimal sighting conditions. And I just had to be free and be there. We were living on Rottnest Island at this time and my Rotto friends helped magnificently with all the details, particularly as I needed to be airborne with Chris and Gail well before school began.

During six seasons of summer research, we steadily gained a clearer picture of the pygmy blue whale population's usage of the Perth Canyon. In order to estimate the actual number of pygmy blue whales, we went back to the system that worked so well for us when the humpback population was smaller: mark-recapture photo-ID. This is a statistical estimation technique that uses a sample of photographs of individuals to compare with a second sample taken a year later. The proportion of resighted whales between seasons is an indication of the overall population size. For example, relatively high numbers of resights between samples mean the sample size itself is fairly close to the actual population size, while low numbers of resights mean that the population is quite large and the samples are inadequate.

Curt and I aimed to photograph as many individuals as possible each season from our small boat platform. Selectively using good quality photo-IDs of the left and right lateral body and tail flukes, 271 sightings revealed that we had seen 208 individual animals. Forty-four whales were resighted within or between years; 31 whales were resighted within the same year and 23 whales were resighted between years. Using six 'capture' sessions (2000 to 2005) and the 208 individuals photographed from the 271 sightings, our mark-recapture model produced a population size estimate of 712 to 1754 individuals — which, for a recovering population with seemingly few threats, was not good news. We had work to do.

After spending hours and hours with these whales, we could really relate to Roger Payne's observation in his 1995 book *Among Whales*:

> With size comes tranquillity. For a whale a passing thunderstorm is but the footfall of an ant, and a full gale an annoying jiggling of its pleasant bed. If you were a whale, all but the grandest things would pass beneath your notice.

SAVING TECHNOLOGY

Whooossshhh!

'The pec fins are forward!' Curt called. 'It's coming up again!'

The turquoise shape glowing in the water just 10 metres from us and almost four times the size of our boat, was a huge 22-metre pygmy blue whale rising to recharge its lungs. The pale grey wall of the whale's body was presented to us for the second time as it brought its near 3-metre pectoral fins forward in the now familiar surfacing pattern preceding a deep dive. This was it!

Pppuwuwwaaauuuffffff! There was a huge explosion of breath from the whale, now right beside us, followed by … silence.

We were expecting to hear a 'twang'. Nick was standing in the bow of the inflatable, a satellite tag still loaded into the crossbow, his mouth agape. We hadn't taken into account that this was the first time he had been so close to a blue whale. Nick Gales is now chief scientist with the Australian Antarctic Division. When he met his first blue whale, he was already one of the world's most experienced satellite tag developers. He had worked with us plenty of times on humpback whale projects, and we had co-developed these tags. However, on this occasion we had forgotten to tell him just how huge a blue whale seems when it is right beside your inflatable boat. The whale didn't give us a second chance that day and we had to bounce and laugh our way back to base at Rottnest Island with the tag still on board. Over the next 10 years we successfully tagged dozens of whales with Nick but we'll always fondly remember this first attempt on a blue whale.

The real trick was to keep the tags on the whales for longer than two weeks and to attach them without causing the animals harm. We were learning where blue whales were travelling within the Perth Canyon complex upwelling/down-welling system, and that they spent about two weeks here, diving to feed at depths of 400 to 500 metres along the shoulders of the deep-water trench.

Revolutionary data from a pygmy blue whale tagged in the Perth Canyon during our Centre for Whale Research and the Australian Antarctic Division collaborative project, revealed it had travelled beyond the Western Australian coastline into the Banda Sea in Indonesia, a journey of approximately 3000 nautical miles. These really are ocean-walloping whales.

Pygmy blue whales have been recorded photographically travelling from the Bonney Upwelling, off the coast of Victoria, to the west coast of Australia. Now the satellite tags filled in the gap and showed their track north to

Indonesia, where a colleague, Benjamin Kahn, also documents pygmy blue whales. What a migration!

To enable the satellite tags to be attached for maximum time and maximum data collection, the shaft of the tags was extended to anchor within the fascia layer of the whales, in the muscle just below the blubber. The life of the batteries was extended by the team at the Australian Antarctic Division by setting the tags to an 18-hour-off/six-hour-on cycle. This long tag measured almost 22 centimetres, an increase from the pen-like 13-centimetre prototypes.

Curt and I worked with our colleagues to learn all we could about these animals and provided the research platform, local knowledge and scientific experience to make these field seasons successful. We aimed to understand more about pygmy blue whales on their northern and southern migrations, as well as learning the subtleties of the northbound and southbound humpback whale migration paths. But we began to worry about the length of these tags and the potential injury to the very animals we cared about, since reports were being tabled about long tags causing festering wounds and thus possibly compromising the health of the tagged individual. Coincidentally, by 2012, many of the projects we had planned in collaboration with the Australian Antarctic Division were cancelled, mostly for lack of funding.

Back in 1993 when we first began biopsy sampling from humpback whales, we used several traditional crossbows. The instructional video for rigging these well-built Canadian weapons indicated they were suitable to bring down any stray moose we might encounter wandering in the forest ... Phew, I felt relieved we would be safe from moose in the Dampier Archipelago — or even the Perth Canyon! Actually, the recommendation might be useful if we ever needed to fend off a particularly interested shark.

Curt was a careful and consistent shooter and together with good positioning of our research boat, we had a 100 per cent success rate collecting a biopsy sample from a chosen humpback whale. For his steady hand and sure shot, we nicknamed him Robin Hood. We also used a crossbow to collect the first biopsy samples in 2000 of those 'blue-like' whales in the Perth Canyon. These vital genetic samples, 3-centimetre plugs of skin and blubber taken from live whales, brought the important understanding that these krill-café visitors were pygmy blue whales.

Working with a pod of whales (one or more animals) we would select an animal for biopsy. If the whale was a singleton then the task of matching the sample to the animal was less taxing. Pairs of pygmy blues or multiple-animal humpback whale pods demanded very precise note-taking. After all, the data is not so useful if it is not assigned to an individual whale. Once an individual was selected, perhaps the focal animal, primary escort, or the lead animal in a pair, and we had checked its gender and behaviour traits, I would drive carefully and steadily behind the whale to gain its confidence. Watching for signs of surfacing by remaining behind the tail flukes, I would speed up at the last moment to get parallel to the dorsal fin area for Curt to take his shot.

The arrow released from the 1.5-metre tautly strung and beautifully arched crossbow had plastic coloured frills for even and balanced flight, while the business end had a screw base for attachment of our carefully sterilised stainless-steel biopsy sample tips, each 0.8 centimetres by 3 centimetres with a tiny fish-hook inside for capturing and retaining the 1-centimetre by 3-centimetre sample. When the arrow hit the animal, a small metal flange at the screw attachment prevented it going deeper into the whale, and the bolt and sample would flop out onto the water. We would immediately locate and retrieve the fluoro-orange floating arrow and quickly transfer the sample into small vials with preservative (depending on the tests), labelling them with details of the animal they had been collected from.

Within a pod of pygmy blue whales we might call them Lead Whale (front) or Follower Whale (back) or we name them after the distinctive shape of the dorsal fin and body markings. A whale with a markedly hooked fin might get the handle Captain Hook on the daily datasheet. Once entered in the computer this becomes a number, but certainly on the day we could tell 'Captain Hook' apart from 'Whitey' — possibly a humpback whale with lots of white on its flanks (a Type 1 or Type 2 pigmentation) — which is important for correctly attributing tail flukes for photo-ID, noting whether a biopsy sample has been collected or a satellite tag has been attached. Getting to know the whales you are working with is very important during any research encounter. We immediately name each animal so that all the data we collect is attributed correctly.

We have used biopsy sampling alongside the all-important photo-ID for many different reasons and in many different projects, all in aid of our primary goal, the conservation of the species. For humpback whales, Howard's year-long Fulbright scholarship study compared the proportion of fluke pigmentation patterns to infer genetic differences and similarities of humpback populations around the world. In 1993, biopsy samples backed up these genetics assumptions. Muriel Brasseur used humpback whale samples we collected in 2003 and 2004 on her behalf to understand the relatedness of 'pulses' of migrating whales for her PhD studies. During the years 2000 to 2005 we employed biopsy sampling to understand just which subspecies of blue whales were using the Perth Canyon. A collaborative project with the Australian Antarctic Division had us sampling humpback whales in Exmouth Gulf during July and August 2007 to compare a range of aspects including migrating whales outside versus inside Exmouth Gulf for Natalie Schmitt's PhD research. During these studies, a Paxarm air rifle was used effectively. Whenever Dr Bob Pitman, world-renowned cetacean observer, is asked to identify a mysterious cetacean species observed at sea, he simply replies 'Remember the Eleventh Commandment' — meaning, 'Never leave home without your biopsy rifle! The truth lies in the genetics.'

The pygmy blue whale data was absolutely fascinating. A week in the life of a pygmy blue whale was literally unfolded with a Whale Lander tag staying on an adult whale for eight days as it travelled north from the Perth Canyon to the Abrolhos Islands, halfway along the Western Australian coastline.

A WEEK IN THE LIFE OF A PYGMY BLUE WHALE

A few years back, Dr Russ Andrews our sat-tagging colleague, joined our pygmy blue whale project in the Perth Canyon to deploy a Whale Lander tag on a pygmy blue whale. The weather was unusually calm and we were enjoying our ride over the deep blue sea. The odd strand of floating *Sargassum* seaweed reminded us of the rougher weather a few days previously. We found a single pygmy blue whale travelling the edges of the canyon wall. It was diving regularly with a 12-minute downtime. Russ prepared the 7-metre ultra-light carbon-fibre pole with the Whale Lander tag attached, Simon readied the Dan Inject gun for biopsy collection, and I checked the batteries of the two GoPro cameras, one on a pole at the stern and the other on my helmet, as well as fussing over my wide-angle camera, making sure any water droplets had been cleaned away. Simon and Russ checked the GoPros on their helmets. Curt's job driving the boat was critical. With a steady manner he read the whales and tracked them, expertly predicting their every move.

Pygmy blue whales swim just below the surface, sometimes 10 or 20 metres down. In the water their 20-metre-plus bodies show as huge bright blue streaks. For five or so minutes we travelled alongside this alluring blue hue. It's a lively, gorgeous colour and in clear calm water, it's a unique spectacle. Ness, our research assistant, once joined us in the inflatable and experienced this beautiful and intoxicating colour.

'Now I know!' she said. Meaning that now she understood the mix of wonderment and adrenalin on our faces each time we returned to *Whale Song* after a blue whale encounter.

This particular whale, Tiny, named for its remarkably small dorsal fin, was very pale grey, and the hue we were tracking through the water was an extremely vibrant turquoise. Watching for the moment of surfacing, preceded by the movement of the almost 3-metre pectoral fins, we were ready.

'Fins are coming forward — hold on!' Curt announced as he accelerated. Instantly we braced ourselves with our legs, since our arms were occupied holding equipment. On board *Orca*, the original tender of our newly acquired *Whale Song*, straddling the pontoon at the stern of this smaller inflatable vessel was the safest option for stability, but I felt as though I was riding a horse at sea! The whale's pecs came forward and the whale was committed to surfacing. The huge rounded head with its flat upper jaw exited the water first, then the blowholes opened, the muscles beneath the fleshy splashguard tensed and the splashguard was raised.

With a strong exhalation, almost 10 metres of the whale's body rolled through the surface beside us. Russ angled the pole downward towards the centre of the body behind the blowholes and successfully slapped on the tag. A moment later Simon took aim, and successfully collected a biopsy sample. As the whale dropped back into the water it rolled slightly towards us, generating a surging wave.

'*Yahoooo! Ughhhhbbbbrrrr!*'

A half-metre wave swept over the pontoon, leaving a refreshing tideline at my armpits. I was awash from a whale and dripping right down to my toes. Yet again, despite being near the stern, I was at the activity end of the boat. *And you know I wouldn't have it any other way!*

○

Having put a Whale Lander tag on Tiny, almost immediately the transmissions began moving up the coastline from the Perth Canyon. The Whale Lander tag is designed to stay on a whale for around 12 to 24 hours. During this time, the whale would show us how and where it was feeding in the canyon and surrounding areas. The tag would slowly work its way out of the blubber and, once dislodged and floating, we would go and retrieve it so more detailed data could be downloaded via data cable. Would we ever be able to catch up with this tag? As the days passed, the journey of this whale became more interesting. With emails zinging across continents to coordinate logistics, our crew member Simon was able to find the floating tag in the flotsam and jetsam of a current-line. This was unbelievable and extremely exciting!

Over the course of the nearly eight days of attachment, this pygmy blue whale made 1677 dives, the average being to a depth of 13 metres. When one calculates the girth of a whale and what would be the most efficient depth for swimming to avoid the effects of the surges and waves at the surface, this whale was travelling just below the threshold of 12 metres. Ninety-four per cent of its time and 99 per cent of its observed migratory dives were within the range of the draught (up to 24 metres) of a large container ship.

The Whale Lander tag (8.9 centimetres in diameter and 6.5 centimetres in height) is anchored to the dermis with three titanium darts that are designed to penetrate only 4.5 centimetres into the whale's tissue.

The Whale Lander tag has revealed some very interesting data. From nearly 1700 dives as the whale travelled over 500 kilometres northwards, feeding dives were identified by reduced body speed as the mouth opened and a jerk of body direction was recorded as the whale twisted to close its mouth to consume the krill. Feeding dives were deep, mostly over 100 metres, and the deepest was 506 metres. The most important detail is that throughout the feeding, migratory or exploratory dives, despite not being at the surface, this pygmy blue whale was at risk of collision from ship strikes. This was most alarming and concerning.

Humans continue to pose very real and present threats to these remarkable leviathans.

EYE TO EYE

'That's it, I am done. I love you all dearly, but I *can* die now ...'

Back on board our small inflatable, camera in hand, I had just emerged from spending two hours with a pair of pygmy blue whales in the endearingly blue and endearingly deep water of the Perth Canyon. My heart was overtaken with the vista of blues.

This slow-moving pair of whales with tall blows in the light air had got our attention from 2 nautical miles away. As we moved towards the pod we collected photo-ID data, carefully noting the dorsal fin shape of each individual and the matching tail fluke. As one whale raised its fluke, the bluer-than-blue Western Australian sky beyond was visible through a tiny hole on one side. That stuck in my mind as a memorable and easily identifiable feature for the animal within this pod and for the future. As we were finishing photographing both left and right lateral body images, we noticed the whales were slowing down and becoming curious. They began to approach our small vessel in closer passes, at first 50 metres away, then 40 metres and then only 20 metres from our starboard side. Moving only very short distances between surfacings, these whales were not exhibiting the migrating, zigzagging or feeding behaviours usual for pygmy blue whales in the Perth Canyon.

'These whales are pretty casual,' Curt said. 'I think you should go in and see what they are up to.'

I needed no second invitation. This is what I live for, but it is not possible every day. Today would be a special day. As researchers, we feel it is important to collect data but only to take opportunities to get closer when the whales present them; we never want to force ourselves on the whales. Within minutes I was gently lowering myself into the water with a minimum of equipment. I had on my summer-weight shortie wetsuit to guard against the 18-degrees-Celsius chill, plus mask, snorkel and fins. Our Nikonos V underwater camera was draped around my neck to record the moment. The water was glassy calm and I was seriously psyched.

I quickly got my bearings. The pair were at the surface, blowing, and with my ears out of the water I heard the whooshing exhalation and inhalation made by the lead animal. Ducking my head back underwater, the scene was both surreal and serene. They were coming our way. Paralleling us just a few metres beneath the surface, a wobbled hue of sparkling blue in the calm conditions, two pygmy blue whales were swimming just a few degrees left of my position, one slightly ahead of the other. The front animal was closer to me, taking my full attention and filling my field of view. The sunlight glinted on the edge of its massive flat round-shaped head; the long silver-toned streamlined body glided gracefully. Barely a muscle rippled and it appeared almost motionless, but it was definitely getting closer. I decided to look carefully at the caudal peduncle, the area from the dorsal fin to the tail flukes. I was keen to observe the relative length of this body region, since a shorter-sized caudal peduncle characterises pygmy blue whales as opposed to Antarctic blue whales, who possess a larger one.

During the previous few years we had seen several hundred pygmy blue whales in the Perth Canyon and we had been concentrating on observing the nuances of identifying them in the field. Given a pair of curious whales, this was a perfect opportunity to make in-water physical observations and record these photographically.

First I saw the tip of the rostrum, only a tiny central point in the middle of the rounded head. This rounded shape was a particularly identifiable feature of pygmy blue whales. The head is composed of a flat, plate-like upper jaw and a basket-shaped lower jaw. From a lateral or side view, the rorqual mouth line begins at the rostrum, paralleling the upper jaw, and then it drops down in an almost comical semi-circular curve, wrapping around and beneath the eye, which is positioned almost a quarter of the way along the body. Lines beneath the mouth line and around the corner of the mouth and near the eyes are called lip grooves. The details of these grooves are simply pretty, radiating forward and aft of the eye region. Ventral pleats originate near the mouth line and extend along the belly to the umbilical, some two-thirds of the body length. With my eyes wide open, I looked through my mask taking in the whole view. The lines all over the body, mouth line, lip grooves and the ventral pleats were engaging, but the spots and blotching patterns were also intensely intriguing. Large pale blue blotches sat just behind the eyes, forward and above the tapered one-metre pectoral fins. They reminded me of the squarish wobbly-edged brown patterns on giraffes, who have similar boxy markings at the base of their necks.

I just floated in the water totally relaxed. My knees were slightly bent and my fins splayed at the surface. From the boat, Curt can always tell my location when I am snorkelling — not only is there the brightly coloured tape on the end of my snorkel, but even when I'm *in* the water, I end up being *on* the water because my derriere is always sticking out. I guess it's good to have something floating! Almost my entire body was relaxed, just in awe as this spectacular animal came closer. The only non-relaxed part of me were my hands and arms. Carefully holding the camera as still as possible, I took frame after frame. Depressing the red rectangular button on the right side of the camera with my pointer finger and using my thumb I wound on the slide film with the lever on the back. I hoped the photos would turn out.

Only a minute or two had passed since I got in the water and we were side by side. The 25-metre whale appeared to halt mid-water and we just hung together eye to eye. I could see the mouth line curved in the characteristic 'near-smile', the heavy brow over the eye was visible, as well as three or four other fetching lines surrounding the eye. As I examined the whale's features and took photos, I saw the grey eye of this whale looking right at me. I continued to lie still, floating on the surface. Was this real? Did this whale stop to look? What was it thinking? Was this a mutual admiration society? I hadn't taken any pictures of the whale's eye, this whole encounter seemed too special. I just wanted to look. Would the moment be ruined if I moved the camera away from my mask, or dived down? After what seemed an eternity but was really only about five minutes, the whale slowly moved on, almost motionlessly beneath me. I looked down on a 25-metre pygmy blue whale. I saw the top of the head, the blow-holes, the mouth curving beneath the eyes, the pectoral fins, the broad flank and all the way back to the tiny, almost insignificantly sized dorsal fin, and then the fine and relatively short tailstock and the gorgeously hydrodynamic tail flukes. A few minutes later there was only one blue left, the bright indigo blue of the Indian Ocean.

I popped my head out of the water, checked that the camera frame numbers had advanced from 20 where I had started, and just hoped like crazy that the photos would do this experience justice.

I plopped back up on the side of the boat, beaming from ear to ear. I was speechless. But only for a moment.

'That whale just stopped and looked at you!'

'Wow, that was incredible — wow!'

I had found words, but only clumsy words. I was still processing what had just happened.

When they next surfaced, the two whales were only 100 metres away, so at low speed Curt paralleled them and I slowly and quietly slipped in the water again without a splash. Moving quietly and confidently, I swam away from *Mega* (our small inflatable), towards the side of the lead animal. Again this whale stopped mid-water and we exchanged gazes. What was this whale thinking? Whales are like dogs and horses — they read people perceptively. If you are comfortable in the water, whales are comfortable around you. Behind me, our research assistants Vanessa Sturrock and Rick Gliddell bobbed in the water as well. Keen to share this unreal experience, we had thought it best to keep things similar to the first encounter for fear of changing the dynamic.

After the whales dived shallowly and we were back on board again with lots of excited chatter, there was a pressing need to change the roll of film in our Nikonos V, the then state-of-the-art underwater camera. While I perched on the pontoon, shivering like a half-set bowl of jelly (I always get cold while swimming, even when the water is 27 to 30 degrees Celsius), Curt thoroughly dried the outside of the camera with a clean dry towel and carefully rewound the film back into the roll. Phew, it looked like the film had been shot properly. In the pre-digital age, to load slide film into a camera meant pulling a small tab out of the roll of film to engage the holes on the edges of the film with the sprockets of the winding gear, which allowed each frame to be exposed properly. Needless to say, an improperly engaged film could ruin your day if you discovered that all those fantastic photos you thought you'd taken, had *not* been taken.

With Curt's dry hands and expert loading skills, I was armed with a new roll of film and ready to go again. Three more times the whales stopped only 100 metres away and I was able to go into the water and continue taking photos, particularly of the lead animal, who appeared the most curious. Three more times this whale 'parked itself' as Curt described it, right beside me as I held my funny half-floating half-sinking position on the surface. Again looking eye to eye. I offered to let Curt see these whales too, but he declined, saying, 'That whale likes whatever you are doing, don't change or do anything different, you go in.'

I will always be grateful to Curt for this beautiful experience. And so, a roll and a half of film later, this whale was immortalised on punchy colour slide transparencies. In 2004, slide film was becoming less popular with the arrival of the digital era, so we had to send it to Melbourne for processing. Usually we waited until we had at least ten rolls to send to make the most of the Express postage. There would be *no* waiting with these two rolls. I had to send them immediately — I wanted to see how the photos turned out!

> As the largest animal, including the biggest dinosaur, that has ever lived on earth you could afford to be gentle, to view life without fear, to play in the dark, to sleep soundly anywhere, whenever and however long you liked, and to greet the world in peace — even to view with bemused curiosity something as weird as a human scuba diver as it bubbles away, encased in all that bizarre gear.
>
> Roger Payne, *Among Whales*, 1995

After two hours and five swims with this pair, Roger knew exactly what I felt. I was at once exhilarated beyond words and absolutely frozen. Shoving down my Vegemite sandwich, barely able to eat properly with my chattering teeth and numb lips, I was on cloud nine.

All the way home — which was at least an hour and a half of bouncing over the waves to get back to Longreach Bay at Rottnest Island — I thought about that whale and that pod. Clearly, that whale was just very relaxed and curious. I didn't think it was making any judgement on my bright mask and fin colours (often called yum-yum yellow in the diving biz). I could only guess what had been the key to piquing its interest. Perhaps being thoroughly relaxed and comfortable in the water, being really aware and comfortable around whales and not pressing our 'friendship' by diving down or doing anything other than my funny half-floating, half-sinking thing. Maybe a combination of these factors gave curiosity the chance to flourish.

Being of rather small stature, just a few inches over 5 feet tall, I could have appeared the size of a small but immobile dolphin. Perhaps my funny floating position was of interest. We will never know but this experience has remained with me for the last 13 years, beautifully etched in my mind. I can see it over and over. Blue whales are not usually curious, so this was an unusual situation and one to be treasured. I believe that interactions between humans and dogs, horses and whales share similar characteristics. It is well known that dogs respond to fear emanating from people, and horses react to humans who are comfortable with them and even mirror people's behaviour. Behavioural research indicates that dogs and horses are able to interpret human facial expressions, since our eyes and mouths are relatively similar. I am not certain the means by which whales assess a passing swimmer, but over several decades of observations it appears that a swimmer who is comfortable in the water can become an object of interest for cetaceans. Of course there will be a range of factors and this may only be one.

A few weeks later I was ecstatic when I saw my self-addressed Express post package in the mail among a pile of bills. It took all my self-control to refrain from opening the slide boxes right there and then at the small counter of the Rottnest Island Post Office. I held my nerve and rode my bike back to Longreach as fast as possible, ran down to the beach helter-skelter and motored our tender *Sousa* swiftly back to *WhaleSong*. Back on board, at our office desk I carefully opened the two boxes of slides.

I was not disappointed. Blue within blue unfolded, slide after slide. I held each transparency up to the light and looked with a photographer's magnifying loop. Each image brought the encounter back. I was eye to eye with that whale. I was wet and in the blue — *I was there.* Yeah, thank you Curt for this. Thank you whales for the opportunity. Thank you Nikon for making such an underwater camera! I looked around, who or what else could I thank? Anything really, I was in a thankful mood!

We pored over the current cetacean photo-ID guides and noted that the pigmentation on the identifying drawings in the guides was not at all like the photographs we had collected. The books didn't show those endearing 'giraffe blotches' and gradients of areas on the head with no blotches above the eye and around the pectoral fins where there were distinct blotches. But these patterns were in the images of the animal I had just swum with. Was this new? It seemed so new. These whales were of the post-whaling generation. Since the late 1980s we had learned firsthand that photography was a fabulous tool, and here it was again, showing us just how much we didn't know about pygmy blue whale anatomy.

The connection with a 25-metre whale, the beauty of these two animals and the serenity of the whole scene had converged in a sublime experience that taught us more about the extraordinary patterns on the body of a pygmy blue whale. This whale had beautifully greeted the world, including a funny-looking snorkeller, in peace.

ANCIENT TRACKS

> As the whale travels through the shifting, changing seas, a map — unlike anything used on land — is learned. The temperature changes; light changes; colours change. The volume of ocean between the surface and the sea floor changes.
>
> *Peter Warshall (1974)*

I blinked and looked again, steadying my gaze. There it was. I carefully lifted my binoculars without altering my field of view and, just like clockwork, again the blow rose into the sky. It was a tall, straight blow and even before it reached its 9-metre height in the light air, the silhouette of an elongated body with an ever-so-small dorsal fin showed at the surface. I grabbed the UHF radio.

'Curt, I've got a whale! It's a really tall blow!'

Curt has graciously learned to respond quick smart when he hears this kind of statement from me. Having got his attention, I handed over the radio to Phil Bouchet, one of the two research assistants with me in the observation tower on the fly bridge on *WhaleSong II*. We needed a vessel with offshore capability and range for our research team to track whales in offshore regions, so in 2006 we spent five months refitting *Genesis*, a 24-metre tuna long-liner, which we relaunched as *WhaleSong II*.

I climbed down from the tower and reached for my camera. Another research assistant quickly radioed the details of the sighting to the wheelhouse where the information was entered on Logger, our computer database. We were travelling on a northward transect on the eastern side of Scott Reef, 200 nautical miles offshore. It was a lovely Kimberley day — the sea was calm, making the colours of the reef even more vibrant. Inside the reef, the water of the lagoon was bright turquoise, indicating shallow waters over the sandy bottom. The orange-brown of the exposed coral at the southern end of the reef was also surrounded on its outer edge by a glowing thin turquoise line that gave way to the glorious indigo hues of the deeper ocean. This scenery on our portside was magnificent and to boot we had whales! But what sort?

In the wheelhouse, Dale Peterson, our first mate, took over the driving from Curt, who joined us on the flybridge. Counting the minutes from the first dive we made notes of the whale's downtime, counted the blows and noted the period of the surface interval. Engaging the steering station to drive from the flybridge, Curt took the controls again and we slowly motored to where I had seen the whale at 000 degrees (a true bearing from my handheld compass) and 1.5 nautical miles away. After 10 minutes we saw another blow, and we were in the mix. After another good look through the binoculars, Curt was puzzled.

'We've got a blue whale here, look at the long body. But where is it going?'

At the time of the sighting we were heading north. From the side profile of the whale, I could tell it was travelling east to west. We knew the two reefs comprising Scott Reef were in that exact direction, so just what was that whale up to? As we approached the pod at the third surfacing for photo-ID images, we realised that there were actually two whales — two pygmy blue whales. I didn't think my eyes could get any larger!

Curt and I weren't certain that the whales were aware of the nearby reef, so we made a quick plan. We would at least get left flank photo-ID images of both animals before they possibly changed course as they approached the rings of exposed reef. Well, the surprises kept coming! With the next surfacing and the next, this pair of pale blue whales continued at a slow but steady pace in a westerly direction. Curt returned to the wheelhouse to check the bathymetry on the chart and carefully monitor our track line in this intriguing encounter.

'They are going between the reefs, they are going through the channel!'

My voice wavered on the UHF radio — I simply could not believe my eyes. From below Curt confirmed what we could see from the flybridge. As we followed behind these pygmy blue whales, 200 metres from the waves breaking on the small sandy patch at Oest Hook (East Hook) on the eastern side of the horseshoe-shaped Scott Reef South, they swam equidistant between it and the ring-shaped reef of Scott Reef North.

These whales had been here before. They knew exactly where they were going. Since everything about this encounter was new for us — pygmy blue whales at Scott Reef and observing them migrating through the reefs using the deep channel — Curt radioed us and said that, rather than trying for right lateral body images, we would hang back and follow slowly right behind the animals. This would serve two purposes. Firstly, we would be able to collect their fluke photographs should they dive between each surfacing. Secondly, by following directly behind the animals and through their footprints, Curt could use *WhaleSong II*'s navigation tracking system, which recorded our route in real-time, to precisely document the migration path these pygmy blue whales had chosen.

Just two years previously we had transformed this 24-metre fishing vessel into a fine whale research expedition vessel with long-range offshore capabilities. The previous skipper had used a range of the latest instruments to measure temperature, depth and current to locate large schools of fish and stay with them for several days. We used these same instruments on *WhaleSong II*'s well-laid-out bridge with every cetacean sighting. Why were these whales where they were? Technology could provide some answers. We hoped that over time with accumulated sightings, we would be able to predict cetacean presence in the same way that the fisherman could predict where to find his fish.

In the middle of the channel between the reefs, Curt came out onto the foredeck for a moment and looked up at the crew on the flybridge perched on the aerial arch. With my camera in hand I wore a grin from ear to ear and just shook my head. This was insane.

'You know, these whales are swimming as straight as a die,' he said. 'Our track line behind them goes in a perfect line. You know what? They are not here by accident and they're not lost. They know the entrance on the east and exactly where the deep water is and where it's shallow.'

One after another, the whales surfaced, one slightly ahead of the other. Was the lead animal teaching the track to the other? Their round, relatively large heads (compared to the heads of the Antarctic blue whales) came to the surface after the enticing turquoise hues of each body heralded their rising. Fifty metres apart two slightly offset blue 'dashes' coloured the dark blue water of the 450-metre deep natural channel between the reefs. A thin layer of water moved backward off the flat upper jaw and moments later the two exposed, A4-sized nares or blowholes opened and each whale exhaled and inhaled. Clouds of water droplets filled the warm air, rising high in fine, swirling movements like the hands of a delicate dancer above each whale. During the five and six blows in each surfacing sequence, the whales only showed their heads, blowholes and the body in front of their tiny stabilising dorsal fins. Counting the blows is imperative as the dorsal fin is often only exposed on the last blow of the sequence. I was poised and ready to get their dorsal fin profiles. As the whales surfaced, exhaling and inhaling with their blowholes, their long, pale blue bodies went on and on, sometimes just slipping under the surface. The arched grey bodies were covered in a pattern of pale grey blotches and in the afternoon light the dimpled texture of the skin was evident.

We continued following and documenting their track. *Their ancient track*. New to our navigation system, but an ancient track to the whales.

Where would they go once they reached the western side of Scott Reef? Within a few hours all was revealed. Passing tiny Sandy Islet, a pretty little sand cay on the western side of South Reef's lagoon, when they were well clear of the left side of the horseshoe of South Reef, the pair made a decided left-hand turn, and began moving southward along the western side of South Reef.

Pretty soon the bright daylight and the warmth of the tropics waned as the sun dipped below the horizon. My regulation uniform of padded vest over a long-sleeved sun shirt, necessary to pad my back and ribs while wrapping myself around the pole of the ship's radar and radar aerial in order to hold still while photographing, was at last useful in the slightly cooler early-evening air. Antique pale pink hues overtook the sky and the sea as we continued to travel with the pair, still so intrigued to see where they were going. Curt moved *WhaleSong II* slightly to their right side for right lateral body photography and we finally bade them farewell when I couldn't usefully push the ISO any more on my camera.

We had spent four and a half hours with these whales and documented their migration path for 10 nautical miles through the middle of Scott Reef. We motored northward and anchored in our designated spot south-west of Scott Reef South. As we checked the photos and completed the data entry, I was so glad we had found those whales. What we didn't know at the time was the importance of this sighting and the significance of this ancient track to our understanding of pygmy blue whales — or the wider consequences.

When we got back within phone range, we called Associate Professor Rob McCauley, our acoustic specialist colleague from Curtin University. He checked his logger (long-term underwater acoustic logging equipment) records and confirmed that he had blue whale calls on the north-east logger, less on the south-east one, as well as more on the south-west one than the north-west one, indicating the flow down the eastern side through the middle and across to the west. This *was* an ancient track. Both acoustics and visual surveys confirmed it. Grinning at the memory of this wonderful sighting, I wondered how many aeons these whales had been travelling this way. I suspected more than we could imagine.

This was in October 2008. Since the beginning of the new millennium we had been researching pygmy blue whales in the Perth Canyon with our colleagues John Bannister, Rob McCauley, Chandra Salgado Kent and Chris Burton. From this unique location, we had also begun an association with researchers at the Australian Antarctic Division using satellite tags to follow the whales' migration. Finding pygmy blue whales inside Scott Reef filled in one more piece of the puzzle of the pygmy blue whales' life history. To realise they were migrating to and from somewhere to the north of Scott Reef was the next mystery. Where were they going and where had they been? This question would consume our every breathing moment. The answer was just a satellite tag away and would be revealed in the following season in our continued collaborative work with lovely colleagues Dr Nick Gales and Dr Mike Double from the Australian Antarctic Division.

TAG AWAY

There was a lot riding on these satellite tags, literally. Development of the tags by the dedicated crew at the Australian Antarctic Division, including the fine craftsmanship of the late Eric King, combined with our fine-scale local knowledge and logistical planning and support for the boat work at each location, led to many successful seasons of research in Exmouth Gulf, the Kimberley and the Perth Canyon.

In working to understand how seismic operations in the tropics or temperate zones affected the way the whales travelled, the Australian Antarctic Division had the chance to perfect a tag that could also be used on feeding humpback whales in the Antarctic. The initial idea of satellite tags that Curt had dreamed up involved a limpet-shaped tag that used a suction-cup with a single small nylon spike. To launch this odd-shaped tag we began by using a toilet plunger from the hardware shop. However, the whales' continual movement and flexing meant the soft suction cups and the spike would only last four to five days — seven at the most.

And so we undertook collaborative research with Dr Nick Gales at the Australian Antarctic Division, the supervising vet for the project. This was the beginning of some major development of the satellite tags. The first tag designs were pen-like in size and shape, cleverly made of antibiotic-impregnated bone cement to reduce the chance of infection for the whale. However, these would not stay attached to the whale for more than two weeks. The Australian Antarctic Division evolved the hardware into longer and larger tags with longer and larger attachment gear. Backward-facing prongs and a floating portion were designed to enter the fascia layer of the whale, just between the muscle and blubber layers.

Once we knew that the pygmy blue whales were feeding in the Perth Canyon, the next step was to understand their movements. Where had they come from? And what other places did they frequent? Having assisted with the deployment of the Australian Antarctic Division's satellite tags on humpback whales, it was a logical step to use them on pygmy blue whales to illuminate their migration routes.

Each time the cases full of carefully sterilised tags arrived in their flash wrappings, we were thrilled about the possibility of documenting the life cycles of pygmy blue whales. Everything we would learn was new.

o

The datasheet from 3 April 2009 on board *WhaleSong II* outlined that just after lunch we had sighted a pod of pygmy blue whales, with six blows observed. This indicated there were perhaps one or two whales. On approach, we noted only one whale, a sub-adult, which we estimated to be approximately 18 to 20 metres long. At 12.45 p.m. we decided to launch our 6-metre tender *Mega* to apply a satellite tag. But with this sub-adult being highly mobile and difficult to approach to apply a tag, this whale was not going to be revealing any information about its migration.

The next day there were whales everywhere in the Perth Canyon. Of the six pods moving between the bathymetry features we named the Wall, the Head of the Canyon, the Gully and the Plateau, we knew that at least one of these six adults and eight sub-adults would travel north with one of our tags. Sure enough, by early afternoon one of the sub-adults had been tagged, and from the Plateau this whale ventured into the annals of marine mammal science. The revolutionary data from this pygmy blue whale and two others tagged in the Perth Canyon that season revealed that they had travelled beyond the Western Australian coastline into the Banda Sea in Indonesia, a journey of approximately 3000 nautical miles. These really are ocean-walloping whales. All the bouncy days were worth it.

We spent long days travelling from Rottnest Island to the Perth Canyon and back, often covering 90 to 92 nautical miles in a single day's voyage. There was lots of time to think and dream while bouncing over the waves for three and a half to four hours at a time. I wrote poems, sang songs and planned the view to capture in my next paintings. We saw odd things. That's the thing: every day you go to sea, you see different things. One day while offshore in the southern part of the canyon, in the very deep water we spied STS *Leeuwin II*, the Fremantle-based 36.5-metre three-masted sail-training vessel, affectionately known as *Leeuwin*. As we moved towards the becalmed ship to catch up with some of the crew, a cetacean pod surfaced between us and this picturesque drifting barque. We estimated the stocky greyish bodies of the six animals to be six to seven metres long, with fine white rostrums and tall, upright dorsal fins. They had us puzzled. Later that week, during our regular aerial survey, I was excited to see them from above, their pale grey bodies contrasting as six bright turquoise dashes in the deep blue water. I even saw them a third time, two years later, once again from the air. We decided they must be beaked whales and sent the images to specialists, but received no definite answer.

Some of the surprises were avian by nature. Every now and then birds took us by surprise. One day a yellow-nosed albatross flew right between the centre console of the Zodiac and the person standing in the bow. This was a very up-close-and-personal view of such a magnificent creature. Sometimes flesh-footed shearwaters tried to steal our pygmy blue whale biopsy samples as we manoeuvred to pick them up after they had exited the whale. And so we would write on the notes, '*NB flesh-footed shearwater DNA could be contaminating this sample!*'

A blustery easterly wind made for another seaward surprise.

'Somebody take a picture, no-one will believe us!' commented Holly Raudino (nee Smith), one of our two research assistants that season, as we all rubbed our eyes at the sight before us. On the water in front of our boat sat a beautiful black swan. What? We surmised that the vigorous early morning easterly winds had blown it off course. We were 28 nautical miles west of Rottnest — this was one very lost bird.

To release the tensions of waiting for good weather, the long trips to and from the canyon, actually finding whales and coping with the odd failures of equipment along the way, the nicely chilled and terribly refreshing evening G and T became an integral part of the daily research routine. I had been a tried and true teetotaller before this work began!

BLUE WHALE FACTS

- A blue whale's heart weighs 2 tonnes and is the size of a small golf cart.
- The heart pumps 10 tonnes of blood through one million miles of blood vessels at a rate of 230 litres per beat, with just 5 to 6 beats per minute.
- The heart has 18 to 20 beats compared with our 120 beats under load.
- The valves in the heart are the size of medium-sized saucepan lids.
- A human head could fit inside the aorta.
- It took six men to drag the heart of a blue whale across the deck of a whaling ship, with oil lubricating the decks.
- A blue whale calf is 6 to 7 metres at birth and gains 114 kilos per day, drinking over 200 litres of its mother's milk per day.
- A blue whale uses 270 to 395 baleen plates to filter up to 5 tonnes of water and krill with each mouthful. It can take 1 tonne of food in its stomach — the equivalent of 8000 hamburgers — and then do it all again two hours later, consuming about 3 tonnes of krill each day!
- The tongue weighs as much as an elephant.
- You could play a game of bridge at a card table for four in the mouth of a blue whale.
- Their lungs pump 2000 litres of air, 500 times more than our lungs.

WHALING THEN AND NOW

Blue whales were highly sought after by the whalers simply due to their large size.

Prior to commercial whaling, it has been calculated that the Antarctic blue whale population numbered 239 000 (202 000 to 311 000 with 95 per cent Confidence Intervals or CI). Research conducted by the International Whaling Commission (IWC) from 1978 to 2001 has estimated the population to be 1700 individuals (860 to 2900), increasing at a rate of 7.3 per cent (1.4 to 11.6 per cent) per annum. Clearly this population has been severely depleted by whaling and remains at a very low level.

The designation of the pygmy blue whale as a subspecies only occurred once the Antarctic blue whale fishery had collapsed and the smaller animals were found in different places. IWC catch data records that 12 618 pygmy blue whales were taken during the period of 1960–1971 in the southern Indian Ocean, indicating that the population was at least that size. A study conducted in 1996 south of Madagascar by the late Peter Best and colleagues estimated the population to be 424 (190 to 930 with 95 per cent CI). In Western Australia, the collaborative aerial surveys revealled an initial population estimate in the Perth Canyon of 30 pygmy blue whales (18 to 49 with 95 per cent CI) and, using photo-ID, a closed time-dependent model estimated the population size as 712 to 1754. With photo-ID matches, genetic studies and sat-tagging pygmy blue whales compared with the Bonney Upwelling, it is most likely the Perth Canyon represents low densities of several hundred to just over a thousand in an open population extending from Tasmania to Indonesia. Pygmy blue whale populations are currently depleted, but not as depleted as the highly prized Antarctic blue whale.

o

In July 1982, the IWC resolved that commercial whaling limits for the 1985–86 season be set to zero.

As a young student I remember hearing biologist and conservationist Roger Payne speaking jubilantly that there was real hope for the whales now and that they would be safe from the onslaught of commercial whaling. This was a big win.

Very soon, Japan, an IWC member nation since 1951, found a loophole in this agreement and each austral summer the slaughter of minke whales in the Antarctic Southern Ocean got underway. Under the guise of 'scientific research', Japan has invoked their right to 'take' minke whales for scientific analysis.

The IWC was set up in 1946 under the *International Convention for the Regulation of Whaling*, and the first meetings were held in 1949. The preamble to the Convention outlines that its purpose is to provide for the proper conservation of whale stocks and thus make possible the orderly development of the whaling industry. The IWC is a voluntary international organisation, and since it is not backed by a treaty, there are major issues with setting and enforcing regulations. Member countries are free to leave the organisation and declare themselves not bound by it. As well, any member state may opt out of any specific IWC regulation by lodging a formal objection to it within 90 days of the regulation coming into force, which is known as Under Objection or Under Reservation. Most importantly, the IWC has no ability to enforce any of its decisions through penalty imposition.

The Convention comprises eleven articles and three protocols. The best-known provision is Article VIII, which states:

> 1. Notwithstanding anything contained in this Convention any Contracting Government may grant to any of its nationals a special permit authorizing that national to kill, take and treat whales for purposes of scientific research subject to such restrictions as to number and subject to such other conditions as the Contracting Government thinks fit, and the killing, taking, and treating of whales in accordance with the provisions of this Article shall be exempt from the operation of this Convention. Each Contracting Government shall report at once to the Commission all

such authorizations which it has granted. Each Contracting Government may at any time revoke any such special permit which it has granted.

2. Any whales taken under these special permits shall so far as practicable be processed and the proceeds shall be dealt with in accordance with directions issued by the Government by which the permit was granted.

Thus Article VIII allows countries to kill whales for scientific research purposes and gives responsibility for regulating these catches to individual governments, not the IWC. Since the zero catch limits were set by the IWC in the 1985–1986 season, several member nations have registered their disapproval, known as Under Objection or Under Reservation, and as such they ignore these zero catch limits regulation. The USSR (now Russia), Japan, Norway and Iceland have taken four species (sperm, fin, Bryde's and minke whales), totalling 25 225 individuals.

Under special permits, four countries — Iceland, South Korea, Japan and Norway — have taken five species of whale (fin, sperm, sei, Bryde's and minke whales) totalling 16 755 animals.

Alarmingly, under the protection of a moratorium on whaling, 41 980 whales have been killed. We, the scientists, the activists, the general public and, most sadly, the whales, have been duped.

The endpoint for these whales was not just the genetics laboratory but included expensive fashionable Tokyo

FACTS FROM BLUE

In May 2014, a 23.3-metre blue whale washed ashore at Woody Point in Newfoundland, Canada. Not wanting to miss such a rare opportunity, a 10-person team from the Royal Ontario Museum was deployed to dissect the carcass, a process which took them five days. It is presumed this animal, a female which the team named Blue, died after being trapped in ice.

Three years later, Blue's 350 bones are being assembled for display at the Royal Ontario Museum in Bloor Street, Toronto. Her skull is 9 metres long and it's estimated she weighed 90 tonnes. Her bones, stripped of blubber, were taken to Trenton, Ontario, in two 18-wheeler trucks. It took nine months to compost the bones in cow manure mixed with sawdust to remove the remaining flesh. Several more months of degreasing followed. This was *not* a small job!

Remarkable statements about the size of a blue whale's heart have abounded over the last few decades. But saying that the heart of a blue whale is the size of a small car, and that a small child could climb through the aorta may well still be true. The museum's Blue is not an Antarctic blue whale, which can grow to 30 metres in length and has been the source for these comparisons. Blue's heart, the scientists at the Royal Ontario Museum have reported, is more like the size of a small golf cart and the aorta the size of a human head. Either way, blue whales are an incredible feat of engineering.

Tragically, nine animals died in the event that killed Blue. At least two others had skull injuries consistent with the ice-crush theory. Blue whales are still rare in the Northern Atlantic, where whaling continued until the 1960s and caused the demise of this population. Worldwide, blue whales number 20 000, down from 300 000 in the 1800s. To date, it is estimated only 200 to 400 individuals inhabit this North Atlantic region. Despite no current whaling pressure, other issues continue to be factors, including ocean contaminants, noise pollution and shipping traffic. From the samples the scientists collected, they are sequencing the entire genome of a blue whale. This is a first.

What a treasure trove of information these scientists are gathering from Blue — her death was not in vain.

sushi bars, as allowed by the Convention. Well, it is a whaling commission. Is it time to rethink the IWC?

In March 2014, the International Court of Justice (ICJ) ruled that Japan's whaling practices did not comply with Article VIII and their catch limits were not justified. Sadly, Japan has ignored the suggestions of the ICJ and recently declared an annual budget of AU$58 million to maintain and refurbish their whaling fleet for the future.

I joined the three–person scientific team on an International Decade of Cetacean Research vessel during the 1993–1994 season. The Antarctic thrilled and delighted me beyond words, but I was horrified knowing this research was all about preparing for the recommencement of commercial whaling. The Japanese are playing a long game. As soon as there is an opportunity to return to full-scale commercial whaling, they will do so. I came home with a fire in my belly to protect whales with long-term data sets and to gain knowledge of where, why and how these populations use critical habitats, which also need protection.

DOLPHINS

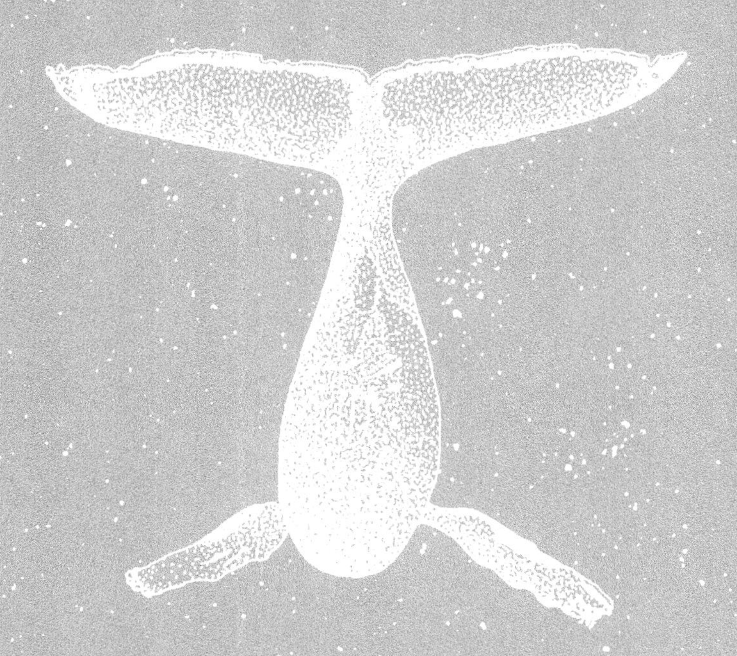

DOLPHINS ON THE BOW!

It had been another day on our catamaran *WhaleSong*, and we were skipping along well with the 20-knot sou'-easterly winds. We were just a little south of the Abrolhos Islands en route to beginning our humpback whale season in the Exmouth region. We were headed north for the warm weather, too. It was May and Fremantle had become chilly. This was the fifth of ten seasons we would spend in the Exmouth area documenting the humpback whale migration at North West Cape with aerial surveys and boat surveys.

Happy that we had sighted some humpback whales migrating north in the sunset, we were on deck relaxing and luxuriating in the warm gold light. It was school term time, the middle of Term II, so Micah, Tas and I had been doing our schoolwork that day. Micah was in Year 6 and Tas in Year 1. Boxes brimming with materials from the state government's long-distance schooling program SIDE (Schools of Isolated and Distance Education) filled the dry bilges of *WhaleSong*. Every two weeks we sent schoolwork to our teacher (several different ones over the course of our SIDE career) at the West Leederville campus for re-marking, and a report on each set of work was returned to us in the mail. Each time we mailed the jam-packed recycled paper packages, I always told the post office staff in a variety of offices along the Western Australian coast, 'A small piece of my brain is in there!'

That day Micah had been working her way through her transport-themed set doing maths, language, society and environment activities, while Tas's language set (a set being 10 days' worth of schoolwork) was centred on flight. Having worked with Micah from Year 1, we were beginning to repeat the sets. Designed for home tutors such as myself to guide children of various ages through the work, the themed material from Kindergarten to Year 3 built on particular concepts and became more detailed as students progressed through the grades. Along the way I learned a huge amount about teaching techniques, including ways to pull my girls up, and I also became a dab hand at making things. With a stash of cardboard boxes and discarded packaging, we could make anything, and I mean anything. I warmed up my scissors each morning and off we went, cutting, trimming and building to our hearts' content. By the end of 13 years of teaching Micah and Tas, we had made cardboard sit-in cars, dioramas of the solar system, native Australian animals, space stations and towns — all manner of *objets d'art*. Spread out on the floor of the cockpit as we sailed along, Tas and I might be creating a helicopter from several Vita-Weat boxes with barbecue skewers and paper for blades as well as designing an airport from which our helo could load and unload passengers and cargo. We even created a picture with the scales of a crocodile made from bread bag clips — we were very creative, it was all go!

We worked from 8 a.m. until 3 p.m., five days a week. Sometimes the girls completed more than an allotted day's work in a day if they were feeling very energetic, just to get ahead. When we were anchored at Tantabiddi, we would swim ten times around *WhaleSong* before lunch. This was at once our daily exercise and our 'shower' in the warmest part of the day. Later, on board *WhaleSong II*, which was twice the size of the catamaran, we only needed to do five laps, but despite now having ample hot water, we still enjoyed our midday swims and saltwater 'showers'.

By 3.30 p.m. each school day we had cleaned up and put away all the bits and pieces. Now the decks were cleared and it was time to relax before dinner. Micah liked to draw, while Tas curled up in a corner devouring a book per day and I splashed around with watercolours. These were our mobile and essential daily chilling activities.

This particular evening, as sunset gave way to twilight and the midnight blue of the night sky, a treat was in store. With the 20-knot breeze pushing us along nicely, the Jenner clan were chatting in the cockpit. Splashes off the aft port deck quickly got our attention. Immediately two dolphins porpoised together beside *WhaleSong*. The beautiful strong bodies of the dolphins shone in the moonlight. These were common bottlenose dolphins, very robust and muscular. As well, tiny ostracods, small crustaceans with bioluminescence, disturbed by the movement of the dolphins, went into an alarm response which appeared as phosphorescence. The swirls of the ostracods, little 2- to 3-millimetre long sea fleas, lit up the darkened water fantastically. Enjoying this beautiful spectacle from deck, we were only metres from these gorgeous dolphins.

Soon the girls grew tired, it was bedtime. 'As you go to sleep,' I told them, 'just think of these gorgeous bottlenose dolphins swimming right beside the hull. They are right next to your cabin, you can see them through the window. Think about those bright swirls in the water …'

With a kiss and a grin, I tucked them into their bunks. I wondered how many children were going to sleep right beside dancing dolphins bathed in phosphorescence beneath a glorious first-quarter moon. How lucky were we?

THE DOLPHINS OF ROEBUCK BAY

When we saw small grey dolphins in Broome's Roebuck Bay at the end of our first season in 1995, we thought they were Indo-Pacific humpback dolphins (then called *Sousa chinensis*), although they were slightly different in appearance from the ones we had seen off Dampier, which the cetacean identification book described as 'rarely found more than a few kilometres from shore, preferring coasts with mangrove swamps' and 'sometimes' entering rivers. As we examined the photos and saw more animals in the field, noting their characteristic 'spitting' of water when spy hopping, we decided the closest species matching those in our book were Irrawaddy dolphins (*Orcaella brevirostris*).

We planned a two-week trip around the creeks and rivers of the Kimberley in June 1996, before our second humpback season with Earthwatch volunteers, to look for these tropical dolphins. Our dear friend and colleague Douglas Elford, photographer at the Western Australia Museum, joined us on board. We travelled to Cockatoo Island, Crocodile Creek, Silver Gull Creek, Koolan Island, Bathurst Island, The Graveyard and then Cape Leveque. The scenery in each of these places was absolutely stunning as we rounded another headland or entered yet another bay.

We only found a couple of these endearing dolphins on this wonderful journey; the largest concentration of these dolphins was right there in Roebuck Bay! This in itself was still data. In science, no data is still data. It certainly doesn't feel good when data collection is absent, difficult to obtain, or patchy, but it is good to keep the big picture in mind. In this case, it appeared that if the animals weren't prevalent in a range of more secluded coastal locations, then the place where they were present, in busy Roebuck Bay, was important. Several years later coastal surveys of these unusual animals by cetacean researcher Dr Deb Thiele and her team revealed that their most-used habitat was indeed in the busiest part of the Kimberley, Roebuck Bay. These detailed surveys provided impetus to suggest reduced speed limits for recreational boats in the area. Genetically different from other species, these dolphins have been correctly identified as Australian snubfin dolphins (*Orcaella heinsohni*) and are integral members of the Kimberley region.

DIAMOND DOLPHINS

At the beginning of my night watch I noticed a note from Sam Wright, our deckhand, on the chart table: *11.05 p.m. Dolphins on the bow saying goodnight through the porthole! Never mind I'll come and tell you at 12.*

We have had some incredibly beautiful dolphin experiences in the last 30 years at sea. With 38 of the 90 species of cetaceans being marine dolphins, the bulk of our individual cetacean sightings on coastal and open ocean routes are delphinids. Some are gregarious and bow ride energetically, like common bottlenose dolphins or Gray's spinner dolphins along the Western Australia coastline. Others are quite skittish and shy, like the dwarf spinner dolphins at Scott Reef or the Indo-Pacific humpback dolphins of the Dampier Archipelago. Exotic species like Fraser's dolphins delight with their infrequent appearances and unusual spade-shaped heads and triangular dorsal fins.

What dolphin species have we tonight? I wondered. Are they still here?

I immediately stepped from the second level wheel-house onto the starboard Portuguese bridge. Sure enough, within seconds I heard the tell-tale sounds of 'phuh-wuh'. We had dolphins surfacing nearby. I was thoroughly blown away with the beautiful sight below. All of the whitewash against our new *Whale Song*'s ice-class steel hull was totally phosphorescent and there were small blobs of brighter ostracod bioluminescence scattered everywhere. This was just as we had seen nine years previously next to our catamaran as Micah and Tas fell asleep. Individual dolphins travelled through the water leaving a blue-white path and as they moved, the outlines of their bodies were visible in the glow.

Quietly entering our cabin, I grabbed my small point-and-shoot camera, and gently woke Curt.

'We have dolphins bow-riding in phosphorescence!'

I was certain that we had only seen this once before in 23 years of research, so I knew he wouldn't mind being woken. The camera struggled in the low light so I immediately put it down. The simple task at hand was to soak in the scene: the foaming and frothing water as it swished away from the hull. Then, from our aft starboard quarter, three dolphins came surging in. As their tail-beats moved up and down, they left long trails of phosphorescence. Their bodies were so clearly visible that I could see one adult animal slightly ahead, and a mother and a juvenile closer and a little aft. The edges of their tail flukes were well outlined and the hydrodynamic shape of their pectoral fins and fine beaks were clearly discernible in the black water. In the diamond light of the bioluminescence I could see we had common dolphins gracing our westward journey past Esperance. They were diamond-encrusted dolphins. Unfortunately, during the few minutes before Curt and Skipper (our darling Jack Russell, who joined our crew in 2006 and loves nothing more than barking at dolphins), came on deck, the animals had moved away.

Pretty soon Sam appeared on deck too. Kindly, he was checking that I had seen both his note and the dolphins. He was worried that I had missed them since he hadn't seen the 10 animals (nine adults and a calf) at the bow for almost an hour. Sam had initially seen them through the porthole of the forward cabin, the bright moving phosphorescence having woken him; next he heard their intriguing whistles. On deck, he could not believe what he was seeing. I was really glad Sam had experienced this. It was an unreal scene. I had hummed and ha'd about waking the whole ship, but since the dolphins had already been bow-riding for quite some time when I saw them, I might have needlessly disturbed *and* disappointed everyone. This is the dilemma of midnight sightings: to wake or not to wake? Having a brief midnight 'party' in the wheelhouse, Resty, Curt, Sam, Skipper and I stayed together for a few minutes, hoping the dolphins might come back. What a treat to be accompanied by fluorescent dolphins by night and a gliding albatross by day. Am I in heaven?

The afternoon before we saw the midnight Diamond Dolphins, the water had appeared 'heavy' or slow-moving when the waves folded at the surface. It was also becoming distinctly greener rather than the bright blue we'd seen thus far, a sign of high productivity. The chlorophyll-a, the green hue we could see in the water, is contained within the cells of surface algae (phytoplankton or plants), which in turn attracts small animal (zooplankton) accumulations. Even in the early evening light we had seen scattered sparkles of those tiny ostracods in the soupy water. *What a wonderful world!*

KILLED BY AN OCTOPUS

Researchers from Murdoch University have observed over 40 octopus handling cases of bottlenose dolphins preparing octopus for consumption, flinging the octopus until the head is detached, and then the arms are broken off into smaller pieces. Gilligan, an adult male observed since 2007, was found dead on Stratham Beach near Bunbury in Western Australia in August 2015 with an octopus protruding from his mouth. The necropsy determined that Gilligan had died by 'non-drowning asphyxiation, specifically suffocation caused by choking'. Sadly the octopus got the better of Gilligan, who the researchers had not observed eating octopus previously.

COMMON GROUND

We were on our way home to Fremantle in January 2012 after collaborating on the substantial research project Behavioural Response of Australian Humpback whales to Seismic Surveys (known as BRAHSS), which utilised several vessels, fixed acoustic arrays, D-tags and 70 personnel based at Peregian Beach in Queensland. From Queensland *Whale Song* had spent six weeks in Sydney (my hometown) as a visiting vessel at the Australian National Maritime Museum in Darling Harbour, where we ran public tours three times a week and hosted several functions with the Friends of the Museum. Loading *Whale Song* with groceries, we bade the museum staff a fond farewell and motored southward once again through the heads of beautiful Sydney Harbour. At the most south-eastern part of Australia, Gabo Island, we turned and made our way along the south coast of Australia.

'Curt, you know in the last two days we have seen at least eight pods of around 120 animals. They're common dolphins, short-beaked. I think you should email Luciana and see if she would like us to collect some samples.'

I was certain we could help our dolphin research colleagues by getting some skin biopsies for their ongoing genetic analysis. Over piping hot vegetable soup at lunch, we made a plan. Curt would email Dr Luciana Moller, a cetacean geneticist at Flinders University in Adelaide, and we would offer our assistance.

Curt came running out on deck to tell us that Lu had replied to his email immediately. In a couple of days' time we would meet up with Dr Kerstin Bilgmann, a post-doctoral genetic researcher working on common dolphins with Lu at Flinders University, and Sue Mason, her field assistant, at Port Weld. This was a great opportunity for Kerstin and Lu to add locations and increase the sample sizes for their research. The premise of this work was to understand the genetic diversity of short-beaked common dolphins across the south coast of Australia, where they are thought to be at risk from local fishing operations.

Common dolphins are relatively small (usually not more than 2.35 metres) and are immediately identifiable with their characteristic tan-to-ochre thoracic patch and brilliant white belly. They are gregarious in pods of 10 to over 10 000 individuals and are often attracted to moving vessels. They actively bow-ride with aerial breaches and leaps. Skipper spotted several of the pods that had us inviting Kerstin on board, literally while he was perched on the back of the couch in the main saloon. Riding beside *Whale Song* and filling the five windows beside the dining table, his excited bark indicated we had dolphins. Sometimes the dolphins were a beautiful addition to our lunch, watching them swimming through the waves abeam while quickly munching. Immediately everyone ran to the bow and the sample collection began. Having spent the previous two decades actively looking for cetaceans to study, this was a whole new way to do research — wait for your subjects to come to you! Skipper became so excited by the presence of these gorgeous animals day after day that we dared not even mention the word dolphin in normal conversation or he would bark incessantly and run helter-skelter to the bow only to be disappointed. To save his little heart, we had to give the animals the code name of 'common mangoes'. But he soon learned what that meant too. Clever Skipper! This was a great chance for *Whale Song* to be a platform of opportunity, just as it would be with dwarf minke whales a few years down the track.

IF YOU'VE GOT AN ITCH ...

One year while sailing north on our *WhaleSong* catamaran with full mainsail and jib pushing us along nicely, we saw splashes a mile out surge towards us. Beneath the splashes the strong, robust bodies of common bottlenose dolphins (*Tursiops truncatus*) appeared and rode the twin bows of our catamaran. Micah and Tas delighted in the gorgeous bodies deftly rolling and twisting. Then they started breaching and leaping, and we saw quite a repertoire of aerial behaviours. One animal began swimming just beneath the curve of one hull.

'Mummy, it's rubbing its belly!' Micah squealed excitedly as we watched the white belly of an adult dolphin actively rubbing against the hull of our yacht — not just once or twice, but constantly. We could feel the pressure of the bang as the dolphin gently hit the boat with its upturned belly. We had just come off the hardstand and had applied fresh anti-fouling paint. I was worried this animal might inadvertently get some new blue streaks — fortunately it didn't!

KILLER WHALE SURPRISE

'Hey, I've got something here. It's a whale or something. It's got a brown body'

Dale, on his 3 to 6 p.m. boat-driving watch, was writing up the ship's log when he looked at the sea beyond *Whale Song*'s bow and caught a glimpse of an animal riding the waves.

Any mention of whales or dolphins gets us on the move from wherever we are. If I am in the galley, in the middle of cooking, I will turn off the elements on the electric stove to make my way to the bow, camera in hand. Sometimes I don't even ditch my apron — it's a good look!

In the golden light of the wintry sunset Curt and I spied what Dale had seen. Four, then five, then six killer whales, resplendent in their black and white colourings, surfed the waves just behind us. Curt repositioned *Whale Song* (our current, ice-class one!) where he estimated they would next surface. Again, we saw their puffy blows on our stern. Turing again, they again kept behind us. Eventually after about four surfacings in this fashion, curiosity must have got the better of this pod. I could imagine them asking each other, *What have we found*? Simon, our second mate, was perched in the bow basket that extended over the side of the boat, and let out a shriek.

The six animals were running fast for the bow! From the two o'clock position they surged towards *Whale Song.* Having tried to get away from us unsuccessfully, it appeared they wanted to investigate us for themselves! Their rotund, smooth black bodies glistened as they porpoised out of the water. The orange hues of the setting sun lit their gorgeous white eye patches a delectable tangerine.

Magnifying the images on my camera using the screen at the back, I thumbed the pages of the whale ID book to check on a couple of features I had noticed. Sure enough, the distinctive shape of the eye patch, its very large area, the prominent cape and the yellowy-brown tone from the presence of diatoms, confirmed these as type B killer whales. Their proximity allowed for good photo-ID images and thus positive species confirmation.

'How was your video, Simon?'

Sheepishly Simon told us it was no good; despite the animals being so close, his wide-angle lens had missed capturing them while he was mildly shrieking with panic! Killer whales 1: Simon 0.

Killer whales are the largest of the dolphins, and with their brilliant black and white markings, are the most easily recognisable cetacean. Long-term studies of killer whales such as the Orca Survey in the Pacific Northwest on San Juan Island, USA, which has extended for over 40 years, has provided invaluable understanding of their acoustics, genetics, habitat, prey preferences and social behaviour. Sadly, the southern resident pods, types J, K and L (fish eaters) studied carefully by Ken Balcomb III and the late Prentice Bloedel II at Orca Survey, are under grave threat. Low numbers of the Chinook salmon they prefer to prey upon are blamed for emaciated animals dying well short of their expected life spans and contributing to poor fertility and increasingly unsuccessful full-term pregnancies. Increased traffic and pollution from large-scale industry within Puget Sound may also be contributing factors.

Killer whale sightings in Australia appear more plentiful, possibly due to several networks, such as Killer Whales Australia, a social media platform where members of the general public can report killer whale sightings. For many years, four or five times a year, killer whales have been sighted in the Exmouth region for several days at a time. Since 2006, these visits have been for longer periods of time as these animals are actively predating on neonate humpback whales. It's wild and woolly and the fights are bloody and to the death. Humpback whale mothers fiercely protect their young but with the killers, the top predators of the ocean, operating as an effective team, frequently the calves are successfully killed.

The Bremer Canyon on the south coast of Australia is another hotspot for killer whales. First noticed by local fisherman and filmmaker Dave Riggs over 10 years ago, killer whales appear every summer and early autumn, congregating in pods of 50 to 100 in the deep water beyond the 200-metre continental shelf at the Bremer Canyon, located 20 kilometres offshore. The bathymetry in that

RECREATIONAL HIGH

Bottlenose dolphins have been documented using the nerve toxin of puffer fish to get a 'high'. The technique of chewing the fish and then passing it on to other animals presumably means each individual receives less than the lethal dose of these dangerous toxins. The animals appear to go into a trance-like state, as if these young animals were 'deliberately getting high'. Filmmaker John Downer carefully and cleverly used spy cameras in fake turtles, fish and squid to capture this unique behaviour in the BBC documentary *Dolphins: Spy in the Pod*.

Cetaceans have so many unusual secrets!

region suggests upwellings and down-wellings (dynamic oceanographic features) drive the process whereby huge amounts of krill, tuna, mackerel, oceanic sharks, giant squid, common dolphins, killer whales, sperm whales and beaked whales congregate for a veritable feast. Sunfish arrive too, as well as many seabirds, including albatross with a 1- to 2-metre wingspan. Occasionally, beaked whales are preyed upon by the killer whales, which have been documented photographically as close to two dozen pods of just over 80 individually photo-ID'd animals as part of Bec Wellard's PhD studies. But it seems the zooplankton and the abundant fish might be the drawcard. Studies currently being conducted on the genetics of the killer whales indicate that these whales share some similar genetic material with the transient (or marine mammal eating) Bigg's whales in the Pacific Northwest.

Bremer Canyon is a vibrant offshore ecosystem worthy of significant protection. Understanding the drivers will be the key to protecting it effectively.

MINKE WHALES

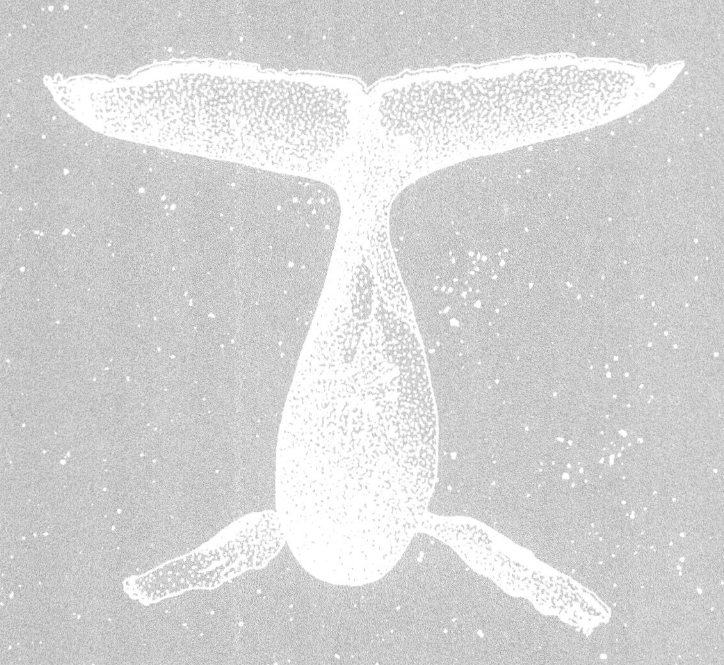

MINKE WHALE MAGIC

'I really love my job!'

This was the polite translation of what Jimmy White, underwater satellite tagger with the Minke Whale Project, yelled out as he bobbed in the water and pumped his fists in exhilaration. On deck we laughed loudly, sharing his sentiments.

A world-first had just occurred: a dwarf minke whale, which our team leader Professor Alastair Birtles had named Spot, had just been satellite-tagged on the right side of its dorsal fin. The Minke Whale Project team had requested topside photographs to document the location and aspect of the attachment, so Pam Osborn, photographer and whale research colleague, and I grabbed our cameras and raced to the Portuguese bridge outside the wheelhouse. This was unbelievably exciting!

Two days previously, at Agincourt Reef (two days' sail north of Hamilton Island) we had met *Underwater Explorer*, a charter vessel for expeditions on the Great Barrier Reef. Anchoring our current *Whale Song* in a sand patch nearby, we had welcomed four members of the Minke Whale Project on board: Professor Alastair Birtles, John 'JR' Rumney, Dr Russ Andrews and Jimmy White. This was to be a superb opportunity for collaboration with our Centre for Whale Research.

We had been heading home from Sydney towards our permanent pen in Fremantle, when Curt called our friend and whale colleague Alastair to see if we could help him out with anything that season. Professor Alastair Birtles is based at James Cook University in Townsville and had co-founded the Minke Whale Project with the late Dr Peter Arnold. The Minke Whale Project's research into dwarf minke whales began in 1995 using in-water photography to collect photo-ID of individuals and to observe their behaviours. Their studies have determined that the lagoons of the Great Barrier Reef are the only known predictable aggregation of dwarf minke whales in the world. Each year they come to the Reef for two or so months, and it is believed their purpose is to socialise. But what they do for the other 10 or 11 months of the year remains a mystery. This pilot study with four satellite LIMPETs (Low Impact Minimal Profile Electronic Tags) from biologist-turned-tag-designer Dr Russ Andrews' workshop would hold the key to finding the answers. For 20 years Alastair and the team had wondered where these gorgeous little whales went after spending two months inside the lagoons.

As it happened, the timing was right and we were absolutely delighted to help out on such an interesting research project. Alastair gave us a latitude and longitude position and a plan was made to rendezvous at 9 a.m.

We were sailing our current boat, the ice-class, purpose-built research vessel *Whale Song*, which had originally been intended for a worldwide research project proposed by Roger Payne's cetacean research group Ocean Alliance. The small ship *Whale Song*, measuring 28.6 metres, was to assist the 30-metre sailing vessel *Odyssey* collect sperm whale biopsy samples. Designed to tow a listening array, it was planned that *Whale Song*, a well-appointed sea-going vessel, would acoustically detect sperm whale calls and direct *Odyssey* to the whales for sampling. The designers, Trinity Yachts, and the builders, Halter Marine, did a beautiful job and the result is an awesome sea vessel that Curt and I are truly privileged to operate.

At Agincourt Reef we anchored just behind the *Underwater Explorer* in a howling south-east wind, tucked inside the edge of the reef near Ribbon 10. Dinghy-loads of computer and camera equipment, snorkelling gear and personnel were loaded aboard *Whale Song*. Our Centre for Whale Research team and four newly arrived Minke Whale Project personnel sat around the table for morning tea, each with a glass of refreshing cool water, to discuss the plans for the next week. I passed around a plate of Pam's delicious shortbread and made sure everyone's glasses were full as we began with a toast from Alastair Birtles.

Raising our glasses brimming with *eau d'Whale Song* — saltwater made into fresh water using reverse-osmosis — we toasted, 'The Minke Goddesses for looking so favourably on everyone.'

At that moment Sam popped his head into the saloon. He hesitated a moment, then burst out with:

'Hey guys, there's a minke whale just 5 metres off the bow!'

'What, 5 metres? Wow!'

Everyone ran to the bow. As we peered over the side, the whale surfaced with a quick 'whoosh' at two o'clock (on the 'ship clock', 12 o'clock is straight ahead and six is behind). It was only 10 metres from the starboard side. The whale was 5 to 6 metres long, slender and dark grey with gorgeous white swirly markings all over the dorsal or upper areas and on the lateral flanks. A relatively tall, curved dorsal fin located about two-thirds along the back from the snout tip appeared in my ninth photo. I had to use my wide-angle lens since the whale was right beside the hull. This was unreal! Even right from the start!

'What's our MO?' I asked Alastair.

'We'll do an orientation regarding the swim-with-dwarf minke whale program and then you can get in the water if you wish.'

Alastair completed the brief orientation and then we were free to get wet. I had already laid out my snorkelling gear in my cabin and in a flash I threw on my swimmers, rashie and shorts and grabbed my camera. Luckily I had a compact waterproof and shockproof point-and-shoot camera that required only minimum preparation. Within a few minutes I was ready. Jimmy uncoiled the 40-metre floating swim-line, attached it to our stern and Simon Kenion (*Whale Song*'s second mate) entered the water. Next Jimmy went in and I followed.

The regulation swim-line for the Great Barrier Reef Marine Park Authority swim-with-whales program was floating white polypropylene with 'hand-hold' rubber inner-tubes attached at 2-metre intervals. The regulations regarding the swim-with permits, such as the one held by JR for dwarf minke whales on the Great Barrier Reef, stipulate that all swimmers must hold the line at all times. Any approaches by dwarf minke whales to the swimmers are thus on the whales' terms, and you just get to enjoy it.

The water was a bright light blue and surprisingly clear given the 25 to 30 knots of wind that had tangled my hair on deck. I imagined the scene that we hoped would unfold: a dark grey body coming towards us out of the blue. Eye to eye with a whale is one of the most intense experiences possible. Hopefully we were in line for this ...

We looked and looked, but alas, no beautiful shape appeared from the blue. Perhaps the time we had taken to get in the water meant we had lost this whale's attention. Now we knew the routine, we had to be ready even faster. Next time, I would endeavour to hold the whale's attention more solidly. I was not sure what this would entail but I was sure that kindly Alastair would advise me.

The next morning at breakfast a cetacean dorsal fin sailed past the window of the saloon.

'Would you please just entertain the whales while we get all the gear ready?' Alastair politely requested. Wow, entertaining whales had just been added to my job description. OK, twist my arm!

I giggled into my snorkel as I took in the beautiful vista of blue water and caught my first glimpse of a curious dwarf minke whale headed towards the line.

o

Alastair had indicated that 20 minutes was the maximum time we should wait for a minke whale to return, so when the whale Sam had sighted didn't reappear after 36 minutes, we decided to lift *Whale Song*'s anchor and get going. We headed north outside the reef towards the Ribbon 5 and Ribbon 10 reefs near Lizard Island. Anxious to hear their calls, we deployed sonobuoys (underwater listening devices) as we moved northwards outside the lagoon, particularly in the channels where we thought dwarf minke whale calls could be detected within the lagoon as we passed. The 'ba-da-da-doing' sound of these animals, also known as the '*Star Wars* whale calls', captured on the sonobuoys and processed through sophisticated equipment such as SDR (Software Defined Radio), filled the wheelhouse. We could hear and see the sounds of these beautiful little whales. Curt decided there and then that their 'ba-da-da-doing' call would be the new ring-tone on his phone.

Alastair was intrigued by the live recordings and watched as the shapes of the signals danced across the custom software on the computer screens. Adding to the beauty of simultaneously seeing and hearing the sounds, the calls were arrayed in the colours of the rainbow that related to a multi-coloured compass rose indicating the bearing of the call and thus its direction. Alastair stood in the wheelhouse grinning, immersed in the sights and sounds. He madly scribbled notes on the timing of each animal's calls and carefully documented the five or so animals that were calling on each sonobuoy.

SEEING WHAT YOU'RE HEARING

A really cool aspect of the software we use on *Whale Song* for whale data collection is that the sound is converted from a VHF (Very High Frequency) radio signal to a moving 'waterfall' graph on screen that shows frequency on the X-axis and time on the Y-axis.

Bearings indicating the direction of the call are represented by a rainbow wheel of colour: a song from a humpback whale shown in yellow tones is from the north, any calls coloured green-yellow are from the north-east, and a purple song is from the west-south-west — you get the idea.

As we have circuited Australia several times, we have amassed quite a collection of cetacean calls that are currently being analysed by several PhD students. This new data will help unlock the answers to a wide range of questions.

We repositioned *Whale Song* at Ribbon 10. As Russ completed the satellite tag preparation, we began to move into the water quietly from the swim platform and shuffle down the floating line holding the inner tubes. Alastair was at the end of the line, next was Curt, then Dale Peterson (*Whale Song*'s first mate), Resty Adenir (*Whale Song*'s chief engineer), Wayne Osborn, Pam Osborn, me, Micah and Tas. The line snaked back and forth as the wind of 35 to 40 knots and wildly surging current moved *Whale Song* side to side with us trailing behind on the 40-metre line. We were pulled to and fro in the surge, but nothing could distract us from the vista. We had five dwarf minke whales swimming around and beneath us — some within 3 metres of us! Holding on to the line with my left hand and taking photos with my right, I was thrilled to see the first animals coming towards us.

Out of the blue, a fine, beautifully streamlined rorqual whale swam slowly and carefully towards the line. The distinctive white band on the pectoral fins showed first in the light blue water as two white 'markers' coming towards you. Then the fine, pointed and darkly toned rostrum, nicely curved mouth and the lines of the throat grooves became clear. Next the shape of the dorsal fin and the scars of the lateral body became evident and then the wonderful white swirling patches of the lower flanks. As the whales became more confident, they came in for closer passes. Soon four whales were coming from every direction.

I was so happy to experience this amazing encounter with my two girls. At first 16-year-old Micah held my left hand, but as the whales came closer, so too did Micah! Squeezing my left hand, Micah then tried to climb on top of me, getting closer-than-close each time a whale swam by. Apparently holding my left hand wasn't enough and pretty soon her entire body enveloped my whole left side. She also captured my free right hand — the one I was trying to take photos with. Quick smart I gave up on photography. I pushed the camera further up my wrist and just enjoyed the scene. I could see I wasn't going to be getting any fantastic *National Geographic* shots with Micah *on* me. It was OK, I could leave that to Wayne and Pam with their extensive in-water experience with sperm whales and humpback whales and high-quality equipment. This whole process was hilarious and an awful lot of fun.

Alastair collected photo-ID of the pigmentation patterns of swirls near each pectoral fin with his underwater camera and furiously made notes of each whale's physical characteristics, including their unique dorsal fins and scarring, as well as the gorgeous individual body patterns and even little goose-necked barnacles on the trailing edges of the tail flukes. With an underwater slate and pencil Alastair expertly drew the whales and recorded their behaviour with each pass, noting the proximity and social activity.

Meanwhile I was receiving a whole-body hug and enjoying the whales. I was convinced that life couldn't get any better! Tas, all of 12 at the time, was more excited and less fazed by the encounter and, luckily for me, did not require body hugs.

As the whales became more confident with our predictable behaviour, they came in even closer. The brown eye with its evident grey ring followed you as it passed, sometimes showing a little bit of white around the edges. They came under, past and towards us. These were truly pretty whales with their ballerina bodies — fine and strong and covered with delicate white and grey patterns. I totally agreed with the description in the Minke Whale Project factsheet that dwarf minke whales are the most highly patterned of all the baleen whales, but I also decided they were the most beautiful, too.

MINKE MISCHIEF

What were the whales doing? What was the reason for their visits every winter? Commercial vessels running the swim-with-whales charters recorded whale sightings in the Great Barrier Reef from April to September each year, but 90 per cent of the sightings are made in June and July. The data collected over the previous 18 years showed that the dwarf minke whales ranged in size from 3.7 metres to 7.0 metres in length. Given the predominance of animals at the smaller end of this range, it appears these wintertime visits are made by adolescent animals displaying socialising and courtship behaviour. They are coming to the Great Barrier Reef during their breeding season. Investigative behaviour accounts for the interest factor with vessels and people, which is totally engaging. In this situation, the continuous 'chu-chunk-chu-chunk' sounds of our anchor chain presumably took the interest of that first whale we saw while we were having our meeting at Agincourt Reef. It is fascinating to work with different species of cetaceans, since each of the 90 different species have their own characteristic physical and behavioural traits. Of course, individuals within a species display variations too and this makes everything even more interesting. As whale biologists, Curt and I were thoroughly fascinated by these whales that are actually interested in people. Who is studying whom?

Most of the whales visited during the peak winter months, but where did they go the other 10 months of the year? This question and others would be addressed with the information gained from the satellite tags Russ had ready for application. This pilot study by the Minke Whale Project aimed to understand the migration of these dwarf minke whales during the other 10 months of the year with this first deployment of satellite tags.

'One of the whales has started spy hopping!' JR called out.

Pam raced out with her camera and caught photos of the whale's head as it bobbed vertically, its fine snout and ventral pleats on show. With his topside camera Alastair caught a couple of baleen claps (when the whale claps its mouth and the baleen is visible), from a well-known whale called Bento, who had been encountered and photographed over seven seasons. Just as everyone was trying to hurriedly grab some lunch, Alastair came into the main saloon and said to Jimmy:

'The game plan has changed. Bento is here. We will stay here and we need to be in the water again!'

Immediately Jimmy leapt back into the water. Here was a good chance of getting a sat-tag away. Alastair held the end of the swim-line, now at our stern. Jimmy was next, then Wayne and me. The water was bluer than blue. The whales circled again; now we had six animals. My view was filled with whales headed towards the line from the starboard side. They crossed beneath us and some headed to the end of the line. Over and over the whales swirled around and past us. Jimmy made a few approaches to deploy the satellite tag using a spear gun, but he felt that the opportunities weren't quite right. Patience was the name of this game. As the afternoon wore on, the anchor chain became interesting again, so Alastair and Jimmy took the bow position with five whales nearby. At around 5.10 p.m., Pam and I were in the main saloon, charging camera batteries when Dale arrived saying, 'They have requested more photographers, they have a sat tag away!'

Out on deck we raced with our topside cameras. These whales had us on the hop! They were the ones directing traffic.

o

Over dinner that night we loosely discussed the next day's plan, knowing of course that it's the whales who actually write the plans.

'If they come to the boat and engage, we stay,' suggested Alastair.

Sure enough, at 6.20 a.m. the next morning, minke whales were circling again. The early morning 'entertainment gang', Tas and I, were enlisted and dutifully held the swim line at the stern. Alastair, Wayne and Jimmy hit the water just after 8 a.m. with all their equipment. The whales, as had been the case the previous day, were most interested in the anchor chain and circled close by the bow. Simon went forward to attach a swim-line with two floats and while he was there, 10 dwarf minke whales cruised around very closely. Again, the anchor chain seemed to engage them, we presumed because it was tinkling, and we relocated the swim team to the bow. Sure enough, over the next few hours (yes, hours!) there were many passes of 10 whales all around the guys. Without much ado, two tags were deployed at 10.10 and 12.05. Jimmy shot the tags with the spear gun right into the centre of each dorsal fin, absolutely perfectly positioned.

The third tag was applied at 1.35. We giggled as we watched Jimmy and Alastair bear-hug in the water, they were delirious and happy and deliriously happy, all at once! Alastair decided to stay in the water to keep observing the behaviour of the animals in the post-tag deployment period.

Jimmy swam towards the back of *Whale Song* and yelled, 'Mich, this is unreal, you should come in!'

'I'll be there in a flash!'

I ran to my cabin to get suited up again. Hovering between 35 and 38 knots, the wind was still howling. Huge waves buffeted the ship's hull, slapping loudly. Once I was in the water, the insanely blue colour of the water and the incredible curiosity of the whales totally overtook my senses. Everywhere I looked there were dwarf minke whales. Fourteen whales swam by slowly and serenely, stacked in four different layers in the water column; all going in different directions. Looking around me, five animals cruised past gracefully. We were a whole bunch of people looking at a whole bunch of whales *and* a whole bunch of whales looking at a whole bunch of people.

Which had come first?

This was the most unreal interactive experience possible. Celebrating the deployment of all the tags, the whole crew came into the water and Jimmy kindly watched us from the bow. What was that pitter-patter sound? Rain? We were getting a freshwater rinse from the top and a saltwater wash from below! I remembered Alastair had commented that there was a high correlation between minke whales and rainbows in winter on the Great Barrier Reef. Yes, it was winter. I could hear the odd bark from Skipper and saw JR and Jimmy on the bow. After a while Skipper's usual incessant barking stopped. I hoped he was OK but realised that since it was pouring, he would be inside, where all good dogs should be! All our *Whale Song* crew was in the water, except Resty, who with Jimmy and JR kept a watchful eye on us as we bobbed in the water. Whales swam by everywhere, ever so slowly — it was absolutely serene.

One of the first things I noticed as the whales swam towards us was their cheeky 'smile' — their curved mouth beneath their fine and perfectly pointed rostrum (not too sharp and not too round) — as well as their eyes looking up with that engaging 'white eye'. Their blowholes were held tightly shut in two dark divergent lines and the area between the blowholes and the dorsal fin was scattered with patterns including grey-black marks and cookie-cutter marks (a reminder of their natural predators) in various stages of healing — some were fresh and white and some were fully healed with pigment returned. Each dorsal fin was unique (some clean, others scarred or bent to the left), the caudal peduncle was fine and muscular and also covered with marks and patterns, and their tail flukes had scratches and scars and thin goose-necked barnacles hanging off the trailing edge. Of course, this was the view from above. From the lateral view, all the different patterns were evident right from the rostral saddle to the peduncle blaze — 10 patterns all up.

These are the most spectacularly patterned baleen whales, said Alastair's briefing notes. Indeed they were. They were outright beautiful.

From my right side I saw another whale head my way. Something about its slow swimming made it appear a little more interested than some of the others. How could this be? We had just spent the previous few days circled continuously by minke whales … How could this one be any different? My mind had already been blown! I turned to face the whale as it slowed to a standstill and began to spy hop: its head broke the surface and its body was vertical in the water as we faced each other. Then it slowly slipped back down into the water, still vertical. I bent my elbows and held my hands close to my face to avoid making any wild movements. As I wiggled my fingers like the tentacles of an anemone, the whale peered at me and I peered back. We were still belly to belly! Its eyes were rotated forward to look right at me, and I could see the white at the outer edge of the eye and even the pink of the blood vessels. We held this pose for a few more moments of stillness. I looked all around the whale's body and saw the beautiful ventral pleats (partly pigmented white and black), which allow the whale to expand and take in water and food for filtering. Still motionless, I noticed a grey scar on the left of the white belly, near the pectoral fin. The whale's pecs were gorgeous, pleasingly proportioned and wonderfully patterned with the characteristic white band of dwarf minke whales. We held this 'dancing' pose for at least several more seconds. It did not matter how long or short it was — I knew I would remember this for the rest of my life.

More whales circled. Here we go, the next round of passes. This was intoxicating and incredibly beautiful. The blueness and the sereneness of a whale spy hopping in your face was remarkable. By this time I was shaking uncontrollably with cold. I had said to myself, 'Just one more pass,' multiple times over the past several hours. Finally, I knew I had to get out of the water. Now, I was *really* cold. I motioned to Simon that I was heading back, and pointed towards the starboard side of *Whale Song*. As I followed the floating line towards the bow and began swimming along the starboard side, the dancing whale came around again and we had a second 'dance' together. These moments of bliss were simply fantastic. Right in front of me, belly to belly, this dwarf minke whale did a spy hop. I was looking into the eyes — no, the soul — of a 5-metre, approximately 5-tonne whale. Yes, a *whale*!

Again, I tried to get to the stern and couldn't stop smiling as I shivered and shook uncontrollably. The laughter just welled up from my toes ... Belly to belly again, the dancing whale spy hopped again. I shook my head and laughed some more. That was three times ...

SAVING WHALES

Curt and I have spent our whole professional lives trying to be careful not to affect the behaviour of the whales we study. Our primary aim is always to protect them through knowledge and understanding. From our years of conducting small boat work with humpback whales, we determined that only every third pod would be curious and 'mug' us. But to have 100 per cent of these dwarf minke whales interested and showing extreme curiosity was unbelievable and very special.

Since the mid-1990s when we started using larger vessels, our research focus has turned to offshore surveys and aerial surveys, so these close passes with whales in-water or from small boats had become less frequent. For the five years (1990–1994) when all our research was done using small boats, whale muggings had been more of a common occurrence, but they were also dearly treasured. Now with our quiet vessel, *Whale Song*, we have been able to have encounters with many different species, but these dwarf minke whales were something else. They wanted to engage and were totally interested. What were they thinking?

Curt and I watched as Simon, Russ and Alastair stayed in the water at the bow. Around 5.20 p.m., the bowline was removed from the anchor chain and all personnel regrouped at the stern. Alastair and Russ stayed in the water near the stern for more than another hour. By this stage I think Russ had been in the water for eight hours and Alastair for 10 hours and five minutes. Were they part fish?

'Wow, Mich, what is up with you?' Wayne asked as we talked over the specialness of this day: 14 whales circling us and a good number spy hopping beside *Whale Song* and even spy hopping with people. I didn't know what was up with me. I only knew that I had absolutely fallen in love with these whales on this crazy day.

'One of the best!' Alastair exclaimed, his hands so wrinkly he struggled to get his wetsuit off. It *was* his second skin, after all! Alastair's wide smile and twinkling eyes told the story. We were all grinning and wide-eyed. I think dwarf minke whales do that to you!

After dinner Russ gave a fascinating presentation regarding the evolution of the design of these satellite tags. Curt and I were most impressed by the care and attention to detail that had gone into this process. With strict protocols in place, the welfare of the animals was the number one priority, a priority reflected in the name of the tags: Low Impact Minimum Profile Electronic Tags.

The four tags that had been successfully deployed in the dorsal fins of four dwarf minke whales, presented the first opportunity for Alastair and his team to understand where these adorable whales go when they leave the Great Barrier Reef. It was a great chance to save these whales. Do they head south to the Antarctic in summer? Are they vulnerable to slaughter by the Japanese minke whaling fleet there?

As I spent time with these animals in the water over four days I could really notice the nuances of each whale. They were individuals in the same way we humans are all individuals although we share similar features: all humans have noses, for example, but consider the vast variety of shapes of noses within a region and across the world. This concept holds true for the individual features of any species of whale, including these beautiful dwarf minke whales. The uniqueness of every individual meant that all of the whales we had encountered in the previous few days could be distinguished from one another by the particular shape of their dorsal fin and their personal body patterning. I loved all of the 10 designated body patterns, including the gorgeous wavy swirls on the flanks and the white markings on their highly mobile pectoral fins.

As everyone downloaded photos and backed-up hard drives that evening, Alastair was grateful for all the high-quality underwater and topside images, particularly those taken by Wayne and Pam Osborn. These photo-IDs would add to the knowledge about individuals such as Bento, who had stayed around for three days and had been returning to the area for seven years.

'Goodbye, whales, and thanks for all the fun, I absolutely loved every minute of it!' I called to the minke whales as they continued to circle, but the sun had set and Curt had brought everyone out of the water. I was so happy that our girls had been with us to experience this. Where would the whales go? And how long would the tags keep working? It was going to be thrilling getting the updates from Russ. This new information would revolutionise our understanding of these wonderful creatures.

These are indeed unique animals. This encounter was thrilling but also sobering, when you consider how vulnerable these animals could be, particularly given their curiosity around humans during the two months they spend socialising inside the Great Barrier Reef. The utmost care has been taken to protect these animals with the swim-with-whales rules, but what happens when they swim south? Minke whales are usually very inconspicuous when migrating, which might serve them well on their northern and southern migration. But where do they feed?

TAGGING MINKES

When we assisted Alastair Birtles and the Minke Whale Project on the Great Barrier Reef in 2013, a Dan Inject tranquilliser gun, the kind used in wildlife parks and game reserves worldwide, was employed for delivery of the tags. These are extremely reliable mechanisms, well balanced for handling with an excellent positioning sight — all pluses in reducing the time spent with whales during an encounter. At Ribbon 10, a Dan Inject gun was initially the chosen tool for airborne deployment of the satellite tags, but since it was so windy, a constant 35 to 38 knots, the team decided on an in-water technique, using a modified spear gun, rather than shooting from the deck.

Using different tags and two different application techniques, this new delivery system had us intrigued.

The LIMPET tags had a variety of settings and could be cleverly pre-programmed as well as reprogrammed remotely post-deployment to change the satellite transmissions for maximum battery life. The data revealed by the tags on the dwarf minke whales was revolutionary and important for their long-term management. It showed that the whales ventured down the east coast and into the Southern Ocean for the 10 to 11 months of the year that they are not in the lagoon of the Great Barrier Reef.

LISTENING IS THE NEW LOOKING

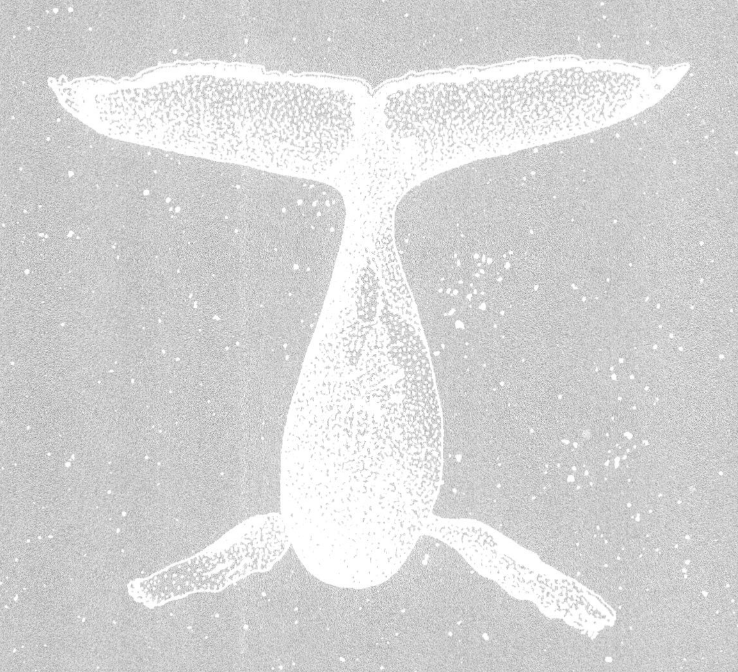

BRINGING SIGHTS AND SOUNDS TOGETHER

'A ship towing an array of hydrophones can effectively hear whales in a strip 150 miles wide — an improvement of seventy-five times,' wrote Roger Payne in his 1995 book *Among Whales.*

Roger has studied whales for many decades and is a true champion for these animals. He and his first wife, Katy Payne, and their four children spent years with humpback whales in Bermuda and Hawaii, and studying southern right whales in Patagonia. Their research put whales on the map, particularly in regard to how and why humpback whales sing. The National Geographic Society produced 10.5 million copies of their record *Songs of the Humpback Whale*. It was a turning point in public awareness of the plight of whales and featured in the Save the Whales campaigns of the 1970s.

Thirty years later, Lou Kipp, a colleague and friend of Roger's, built *Whale Song*, the custom-designed research vessel that Curt and I are now proud to own. During construction, the exact same equipment used to isolate and silence motors in US nuclear submarines was installed throughout the engine room of *Whale Song*. Consequently, she is a very quiet boat.

In 2011 we brought *Whale Song* home from Malta to Fremantle via a wonderful five-month journey around West Africa. We had suspected she might have been a 'quiet vessel', as some of the boat plans were labelled 'mil spec' (military specification), and our experience with a pod of beaked whales we encountered in the Indian Ocean seemed to confirm it. Our friend Associate Professor Rob McCauley of Curtin University joined us on the Cape Town to Mauritius leg of the journey. East of the African continent, at the Madagascar Plateau, the breach of a cetacean 2 nautical miles away caught my attention. It turned out to be a pod of at least six strap-toothed whales. These 6-metre long, spindle-shaped beaked whales are described as 'difficult to approach' and 'have been observed responding to ships by slowly sinking below the surface'. This pod of mixed age and gender (as beaked whale expert Bob Pitman later determined from my photographs) swam around and around *Whale Song*. They remained at the surface and behaved totally differently from the descriptions in the whale guides.

Not long after our return to Fremantle we were in the Perth Canyon with a pygmy blue whale near the bow. While Curt expertly drove *Whale Song*, I made behavioural notes and collected photo-ID. We'd deployed a sonobuoy and as I entered the main saloon, the calls of a pygmy blue whale filled the room and danced across the screen.

Specifically, these were the sounds of the pygmy blue whale swimming right in front of us, and it was louder than our ship's engines; this in itself was not unusual, but engines that were so very quiet were.

A quiet boat where only Skipper and I were the noisy components — well, this was a cetacean biologist's dream.

O

For many years visual surveys have been the primary method of determining presence or absence of cetacean species across regions, using transect lines to maximise coverage. Curt and I have employed this technique for almost all of our career researching whales. From 1987 to the present, we have spent every day at sea scanning the horizon for marine mammal cues: the blows at the surface (exhalations and inhalations) like columns of white smoke rising up from the water, or brief glimpses of bodies or splashes, or exuberant surface active behaviours like breaching, which humpback whales often display.

With this new technology, however, there was the possibility of listening for whales, rather than simply looking for them. One day as we were working offshore, our colleague Chris 'Daffy' Donald was monitoring a sonobuoy. He came into the wheelhouse and informed us that a humpback whale was singing at 180 degrees true, 1.5 nautical miles away. Sure enough, a few minutes after Curt turned *Whale Song* to the south, a humpback whale surfaced right where Daffy had said it would be.

'Daffy, that is cheating! We have been looking for years — now we are listening and then finding!'

This was revolutionary, but also worrying, especially for me. My eyesight had always been useful when it came to whale-spotting. Had I lost my job? As it turned out, my concern was short-lived because it is still paramount for biological acoustic research to actually *see* the animal making the noises you are hearing on the hydrophone or underwater microphone.

o

One of our first expeditions with our new vessel, working with the Australian Government Department of Defence, was to travel from Fremantle to Darwin, and we set off on 13 August 2012 to make the approximately 2000-nautical-mile journey north up the Western Australian coast. The nine personnel on board included the *Whale Song* crew of five (Curt, Dale, Resty, Sam and me), Daffy (Defence), Gary (L3 Oceania), Maria Garcia Rojas (CWR post-doctorate position) and Inday Ford (CWR research assistant). Skipper, of course, was also on board and appeared mildly excited about our departure, but over the next few days he became increasingly glum without Tas, his cuddle-buddy. He was depressed, poor little soul, but snuggling next to Daffy seemed to help. Tas, then 12 years old, had just started boarding at St Mary's Anglican Girls' School in Perth. Micah had already finished her secondary schooling but had thoroughly primed Tas for successful boarding school life. Sailing over the horizon away from my two precious girls felt the strangest thing possible. This was also the first time in 13 years that I was not doing schoolwork with my girls. I promised if everyone was *very* good, I would reward them with a round of gold stars — I still had packets of them ready to go!

Our colleagues at Defence, Daffy and Dr Doug Cato, took the opportunity to collect invaluable data with the acoustic equipment on board. And so we developed the Western Australian Acoustic Profile (WAAP), a real-time ambient noise and cetacean survey along the Western Australian coast following the 100-metre contour with acoustic and visual observations made four times daily.

We began the first of many sonobuoy deployments on the first day. Sonobuoys are essentially underwater microphones that use radio frequencies to transmit the underwater sounds. Once deployed in the water, within a few minutes the radio signal is retrieved on board with sophisticated hardware and processed with even more sophisticated software, all created by the gentlemanly bright sparks at L3 Oceania and Sonartech Atlas. This meant that real-time whale songs were being piped into the wheelhouse as the digital signal was recorded on the hard drives.

Listening to this live 'ocean music' for up to 10 hours each day was a beautiful experience. *Whale Song* was indeed living up to her name. The wonderful sounds of the sea filled the whole vessel. The chirps of dolphins and whoops and trills of the north- then southbound humpback whales filled my sleeping hours between 3 a.m. and 7 a.m. after my midnight to 3 a.m. watch when the 'Dawn Patrol' was collecting data. It was great to dream, nestled in a cocoon of whale songs.

By Day 3, we had fallen into a good routine, working out the nuts and bolts of our regular ship's watches, the cetacean observation schedules for the three observers (Maria, Inday and me), and Daffy and Gary dividing up the four takes per day of running the recording systems. Over 11 days we travelled 1985 nautical miles, deployed several hundred sonobuoys, and collected around 30 GB of data.

Some of the highlights included large numbers of humpback whales playing and belting out tunes off Ningaloo Reef; a sunset feeding frenzy with three Omura's whales; five huge oceanic manta rays barrel-rolling at the surface with krill leaping skyward to escape from them; and a lovely quiet pod of spinner dolphins just swimming gently through the krill soup. We encountered another frenzy featuring two more Omura's whales and a group of 50 to 80 false killer whales, which had a nursery group of smaller individuals bordered by the larger animals out wide. We had a 'bong' fish, something that went 'bong, bong, bong, bong, bong' five times with great resonance (most likely an Antarctic minke whale); 'carpenter' fish chipping away (sperm whales) and a Cuvier's beaked whale made impressive 'zzzzipppphhh' sounds. The sights at sea always amaze me — panicked flying fish, squawking crested terns, bobbing jellyfish, dancing Wilson's storm petrels, red-footed boobies perched atop our foremast — but now bringing the sounds and sights together, we had indeed entered another realm.

On Day 11 we arrived safe and sound in Darwin Harbour, feeling thoroughly thrilled with the data collected. The Western Australian Acoustic Profile was a success. Curt and I felt very privileged to be doing this work, and were elated that our truly quiet vessel had been a recorder and defender of the sounds of the sea!

DOLPHINS AND DYNAMITE

In late August 2012 we sailed to Scott Reef, 180 nautical miles north-west of Broome, to listen for cetaceans for a week in and around this special offshore reef system. Scott Reef is one of several in the tropical deep water of the offshore Kimberley that includes Cartier Reef, Ashmore Reef, Seringapatam Reef and Rowley Shoals. Having encountered a wide variety of cetaceans when we visited Scott Reef and the Browse Basin in 2008 and 2009, we were keen to use the high-quality acoustic equipment on board *Whale Song* to acoustically detect the likes of pygmy blue whales, false killer whales, Risso's dolphins and dwarf spinner dolphins.

Scott Reef actually comprises two reefs. North Reef is ring-shaped and has only two narrow entrances to its sandy-bottomed lagoon, which reaches a maximum depth of 11 metres. South Reef is horseshoe-shaped, with the open portion facing North Reef. The deepest water in South Reef is 50 metres, while the 1600-metre wide channel between the two reefs extends down to 450 metres.

(This was the channel where we followed the pygmy blue whales in *WhaleSong II* some years before, when we discovered that this was an ancient track of theirs.)

Over the course of the next few days an alarming trend unfolded. We began hearing and seeing blasts. Our colleague Rob McCauley had told us to watch out for dynamite fishing, an illegal activity carried out by Indonesian fishermen. Rob was absolutely right. On the screen, in real time, we could see single and multiple explosions several minutes apart, audible as quick blasts. For the first few nights we sat in the wheelhouse listening and looking at the screens with big eyes. Curt took screen-grabs on the computer and recorded the time and type of each event. We informed the appropriate departments of our findings and Customs vessels began investigations of the 17 vessels in the area.

On the fourth night an incredible story played out on the screen in front of us. We heard cetacean sounds that we believed were dolphins or perhaps beaked whales. These animals were hunting using a particular frequency that sounded like cicadas en masse, and this filled the screen in an arc, quieter at first, then louder, then softer again. Next came an individual '*pssck*' call and the frequency changed as the animals were whistling and chatting near the surface. After five to eight minutes there was another individual '*pssck*' call and the pod began the cicada hunting sounds again. We watched this hunting-chatting routine occur several times, marvelling at how the leader appeared to give the order for the pod to move as one. Sometimes during the hunting sessions there was a little whistling chatter, presumably of mums with calves at the surface, those that couldn't stay down too long. After two hunting sessions and two of the whistle-chatter sessions the pod was back hunting again. In the middle of this we heard and saw a bang, a blast that appeared as a solid white line across the computer screen. It was a broadband detection crossing all frequencies being recorded on that sonobuoy. Curt and I had eyes like dinner plates as we looked at each other. What would happen now?

After three minutes, the whistle and chatter came back, but it was shorter than the previous surface sessions, then for at least the next 30 minutes there was silence. These animals had left the area. What did they communicate in that last chatter together? I can only presume something like, 'OK, stay together. Is everyone here? Let's go.' What, if any, were the casualties to these animals? We were stunned, shocked and saddened that the reef and all her inhabitants were being affected by illegal fishing practices employed by supposed traditional fishermen. Having alerted the customs officials days earlier, we hoped there would be a good outcome for our 'reefy' friends, great and small.

o

On our way home to Fremantle, six days after leaving Scott Reef, we pulled into the lagoon at Tantabiddi at Ningaloo Reef for an overnight break. As I snorkelled around the beautiful coral bommies in the sanctuary zone (where I used to take Micah and Tas when they were really little), a Department of Environment and Conservation vessel approached.

'The reef looks good here, well done! It is much better than Scott Reef ...' I yelled to the rangers from the water, my mask perched atop my head.

One of the rangers replied, 'Have you heard? They caught some illegal fishermen at Scott Reef yesterday.'

We couldn't believe it. The rangers told us how the Customs officers we had informed about the blasts had eventually found evidence of illegal fishing at the reef. They had found cordite (wicks for explosives) and jars of dynamite ditched beneath one of the traditional wooden fishing vessels anchored in the lagoon of South Reef.

The fishermen were prosecuted by the Australian authorities, and the evidence of the explosives and the sound data we had collected resulted in them being fined and their vessels burned. I feel sad that people have potentially lost their family incomes, but this fishing practice has been at the expense of the health of our Australian reefs. It is important, however, to recognise that not all Indonesian vessels break these laws. Nevertheless, it was gratifying to see the determination of Australian authorities to send a very strong message to those who used damaging non-traditional fishing methods like dynamite.

THE FUTURE IS IN LISTENING

'Flybridge — this is the wheelhouse. Hey, we have "carpenter fish" — sperm whales. The bearing is to the south, see if you can see the blows. The signals are strong, the animals must be close!'

From the flybridge, the top level of *Whale Song*, the cetacean observation team were looking for marine mammals in the waters just to the west of Scott Reef's South Reef. We used polarising glasses and binoculars to scan the turquoise waters inside the lagoon and the darker, deep blue seas out to the west, our cameras ready to capture any of the animals we encountered. To date humpback whales had been seen breaching in the lagoon and indeed signals of humpback whale songs appeared on the computer screens linked to the sound gear. Bryde's whale 'purrs' had also been recorded on the sophisticated sound equipment, as well as the distinctive low-frequency booms of blue whales.

As soon as the flybridge observation team had the word that there were whales to the south, Curt brought the vessel around and we travelled directly towards the hammering sounds that submariners in the 1970s named 'carpenter fish'. Within only five minutes more excited news came from the flybridge.

'Hey, there they are! See the blows — four ahead and a fifth out to the west another half a mile. OK, we are on the way!'

Sure enough the whole crew was delighted to see the exhalations of the sperm whales who had been responsible for the noise on our equipment. They were slowly cruising northwards in the deep water alongside the reef. These animals have characteristically dark, wrinkled skin, which appears mainly dusky brown. The size of these nearly 12-metre long whales, and that they were travelling together within a 2-kilometre area, indicated that they were sub-adult males in a 'bachelor pod'. Strong, left-angled blows characteristic of sperm whales bellowed skyward above the calm tropical waters and the rounded or triangular hump of the dorsal showed clearly, as did the exaggerated 'knuckles' on the tailstock when they lifted their mighty flukes in a sounding dive.

Visual identification and cataloguing of animals is effective, but the addition of the acoustic component enables us to expand our scientific horizons even further. Collecting cetacean acoustic data can be used for individual acoustic profiling and tracking, habitat usage and even seasonal and temporal distribution of particular species.

Deploying a sonobuoy opens a window on the world below, revealing an ocean full of noise. Combining visual encounters and photo-ID with the acoustic profile of a cetacean is a new aspect of an old science. Listening in to the world below can also help determine the extent of human impacts on cetacean habitats.

o

The high-quality acoustic records we have been able to produce with *Whale Song* have added significantly to scientific knowledge, particularly for infrequently observed species such as dwarf sperm whales. The data we obtained of the acoustic signature of a dwarf sperm whale we encountered near Rowley Shoals contributed to an extensive scientific paper describing its acoustic features, and may well be a world-first recording. In this paper, other cetacean species represented by acoustic spectrograms are new recordings of pygmy killer whales, striped dolphins, Omura's whales and Fraser's dolphins.

One aspect which absolutely fascinates me is the unique frequencies that each cetacean species transmits. The individual characteristics of their calls or songs are immediately recognisable. The frequency of a call represents the number of times per second compressions and rarefactions of the water molecules occur as the sound passes through the water. Both fin whales and blue whales occupy the lower end of the frequency scale with calls at 20 hertz, meaning there are 20 cycles per second of compressions and rarefactions.

The lowest audible frequency for humans is 20 hertz and the highest is 20 000 hertz. Southern right whales emit sounds at 30 to 500 hertz, while Antarctic minke whales make a call known as the 'bio-duck' at 60 to 140 hertz.

Common dolphins produce clicks between 3 and 24 kilohertz and Cuvier's beaked whales make sounds of 13 to 64 kilohertz. How did this come to be?

Looking at the physics of sound production, the wave size (and thus frequency production) may be relative to the physical size of the animal producing the sound. A large baleen whale, for example, produces a low-frequency call capable of transmission across ocean basins. Smaller cetaceans often make higher-frequency squeaks and trills. Despite knowing the physical characteristics linking the caller and the call, I like to imagine these differences evolved as a matter of courtesy, as if a blue whale, for example, had said to a common dolphin:

'No, you take the high, I'll take the low tones.'

Decades of research has shattered the public perception of dolphins as symbols of peace. They have combative natures when it comes to age-old behaviours such as males competing for breeding opportunities with females — the teeth-marks on the bodies of dolphins are the tell-tale signs of battles won or lost.

Similarly, what the ocean sounds like is also generally misunderstood. It's not quiet, as many think. While snorkelling or diving in the ocean you may be treated to a cacophony of sounds: snapping shrimp make deafening crackling noises when individuals snap their large claws, possibly in territorial stoushes; dolphins communicate with whistles as they travel together, roaming the coastal seas and open oceans; and male humpback whales sing intricate songs on the breeding grounds as they try to attract as many females as possible each winter breeding season.

There is a wide range of sources for the sounds filling the aquatic soundscape. Natural sounds are the geophysical events such as earthquakes, icebergs calving, and weather events such as rain. Biological sounds are the vocalisations of the animals (crustaceans, fish and marine mammals). And man-made sounds include ocean floor–based equipment for oil and gas installations, shipping and seismic exploration.

Worryingly, worldwide the ocean soundscape is getting louder. Over recent years, the noises in the 20-hertz range (the frequency emitted by ships) are increasing. Fin whales and pygmy blue whales characteristically vocalise in this range. Alarmingly, Rob McCauley's long-term acoustic loggers that rest benignly on the seafloor for up to a year at a time have recorded that the frequency of the call produced by pygmy blue whales has decreased by 1 hertz over nine years. Are these blue whales changing their call to ensure they will be heard by their conspecifics? Is this similar to a person having to yell in the bar as the number of patrons swells and the ambient noise levels rise throughout the evening? Are we causing pygmy blue whales to change?

What legacy are we humans leaving if we are causing these ancient animals to alter their basic behaviours — traits evolved for their very survival?

Applying slippery paints to ships' propellers to reduce marine fouling and thus reduce the cavitation that creates noise-producing bubbles — and thus noisy ships — should be the way of the future. Interestingly, the Global Financial Crisis may have been good for whales. With less cash around, corporations and businesses slowed the speed of shipping. Across the board there were fewer customers for goods, so travelling slower became more economical. This was noticeable in our roles as master mariners travelling the high seas of the world, particularly when we brought *Whale Song* the 13 000 nautical miles home from Malta to Fremantle. Slow ships are cheaper to run and, importantly, are quieter ships. We are certain that quiet is *better*.

Taking major shipping through the Great Barrier Reef is unwise and potentially damaging in a multitude of ways. Increasing the chance of a collision harming the largest living organism on Earth is bad enough, but risking successful recruitment of larval fish and many other marine animals by reducing settlement on substrates by sessile coral reef ecosystem inhabitants, foraging and as well mating opportunities, by clouding the soundscape is simply not worth it.

IT'S UP TO US

On a global scale we humans need to be clever enough to fix the problems we have created. Climate change as a result of the world's industrialisation has been significant. Whether you believe it or not, the evidence is there. The Arctic and the Antarctic ice caps *are* melting. Weather events are different, more extreme. During any one news cycle, journalists can be reporting on devastating floods, life-threatening bushfires and raging cyclonic winds, and these news stories could just be about Australia. And by the way, increasing global water temperatures are killing the Great Barrier Reef.

Reducing waste and slowing down our increasingly fast-paced world has never been more necessary. Our ready access to information and services has, sadly, made us less tolerant, less patient and less able to concentrate. The resources of our planet Earth are finite and it is possible that in the near future they may no longer be able to support our constant desire for progress and development.

On a global scale, given the will, we can make a difference. There are various things you can do. Shop locally, keep chooks, ride your bike, go solar, re-use, reduce and recycle, use the sun, wind and rain and love whales. For our children and our children's children, we need to care for our oceans and our earth.

We all need to step out into the wild and reconnect — the *real* wild. Be sure to take your children with you. You'll be refreshed.

COOL WATER: ANTARCTICA

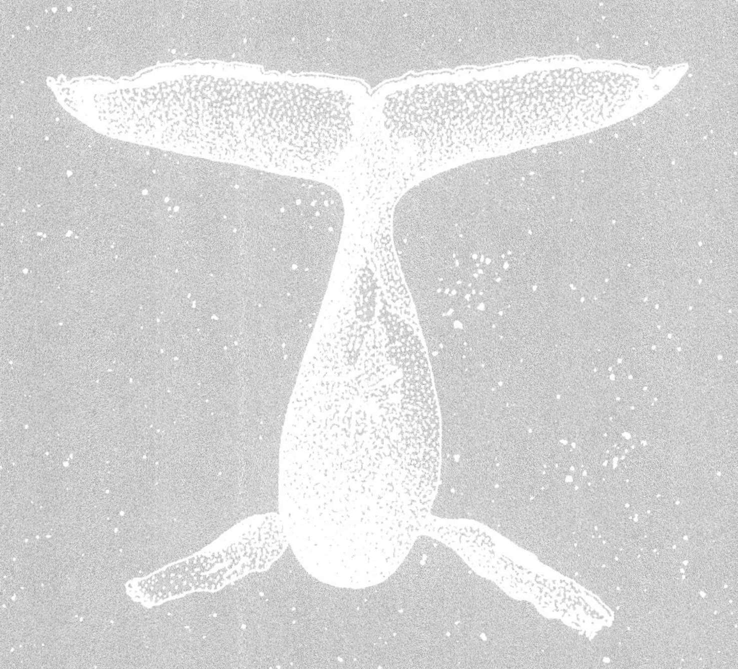

TWENTY YEARS IN THE MAKING

As I looked down into the calm dark grey water just beneath our bow, I could see an adult humpback whale was about to surface at any moment. With a split-second roll of its body, the scene was filled with lines of pink, white and black, close to 10 metres long. These lines were the three dozen ventral pleats or throat grooves on the belly of this whale, extending from the jaw plate to the umbilical, like a piano accordion expanding in and out to allow up to 2 tonnes of water to be filtered at a time. The belly extended enormously with each filtering mouthful. Stirring in the cool water beside our vessel, two 40 000-kilogram humpback whales quietly broke the surface, their mouths agape and almost hinged sideways as they fed on krill, *Euphausia superba* — the cornerstone species of the rich Antarctic ecosystem. Then, the almost unhinged upper jaw with its 270 to 400 black to olive baleen plates made of keratin (just like human fingernails and hair) came into the view.

Towards the centre, these fringing bristles — effectively the filter-feeding screen of these 'moustached whales' — were lighter coloured. When I presented talks on whales at the Pilbara Camp School in the early 1990s, I was the 'Whale Lady', and to the delight of the children I described Curt as the 'Whale Man', since he had a moustache like a filter-feeding humpback whale. Often I heard whisperings of 'That's the Whale Man, look at his baleen!' Well, here we were in the Antarctic, watching humpback whales using their baleen plates!

I blinked hard, clearing away the watery tears caused by the chilly Antarctic air, and fired off at least 50 photographs of the cetacean action in the water below my feet. This was unreal. The whales were active, *really* active, and they were *feeding*. I blinked again to make sure I was actually seeing these humpback whales and taking in the entire scenario.

Was I really here? Yes.

Are we really in the Antarctic? Yes.

Are we actually on board our good ship *Whale Song*? Yes.

Now, is my camera lens clean and the battery still good? Yes.

Yes, all systems were ready for deploying a satellite tag and collecting a biopsy sample from one of these humpback whales.

'OK, Russ, here we go!'

For decades Curt and I had yearned to study humpback whales in their southern polar feeding grounds, complementing our several decades of research in their northern tropical breeding grounds. We had both been south before, but to go together on our own vessel — this was an expedition we had been dreaming of for a very long time. After 23 years, four boats (the 5.3-metre *Nova*, the 13-metre catamaran *WhaleSong*, the repurposed 24-metre fishing vessel *WhaleSong II* and now our ice-class 28-metre *Whale Song*), and a good dose of sheer determination, we had got here. We were in the Antarctic. Even now, several years later, I can still feel the thrill and the excitement.

o

Way back in 1990, when we were bobbing around the Dampier Archipelago in *Nova*, we witnessed the humpback whales' irresistible drive to migrate north to the warm tropical winter breeding grounds. Their southbound journey to feed in the cool Antarctic during the austral summer was also an essential part of their life story. Humpback whales are two-speed whales — they are either feeding or breeding. While working with humpback whales during their winter breeding season, we hankered to document the different behavioural facets, seasonality and end-point destinations of their extensive annual migration. Our dream took hold: we aimed to document humpback whales in their northern winter 'bedroom' in the Kimberley, Western Australia, as well as their southern summer 'kitchen' in the Antarctic.

Baleen whales (Antarctic blue, pygmy blue, fin, sei, southern right, minke and humpback whales), who have baleen rather than teeth, undertake enormously long migrations on a yearly basis. Humpback whales are known to make the longest seasonal migrations in the animal kingdom. In the Southern Hemisphere, annual movements from polar feeding grounds to equatorial breeding grounds along the coasts of Australia, South America and South Africa may be undertaken to reduce the predatory pressure of killer whales during the breeding season due to their reduced numbers in the tropics. Another theory proposes calm water as the driver for seasonal migrations. Resource tracking, whereby reduced fasting is offset with 'stop-over sites' for foraging and physiological maintenance, or the process of accelerating skin regeneration in warmer waters for killer whales, or even reducing barnacle adherence in the warmer waters for humpback whales, could all be valid explanations.

The northern migration of humpback whales is further offshore, as the animals make a beeline for their northern destination. The southern migration is further inshore, almost winding its way down the coast. The cows with their young calves migrating south stop along the way to recoup in 'resting areas' — calm, shallow bays and gulfs. The males, aiming for as many mating opportunities as possible during the winter breeding season, follow the cows and calves inshore into places including Nickol Bay, Exmouth Gulf, Shark Bay, near Rottnest Island and south-west coastal bays including Geographe Bay and Flinders Bay at Augusta. All of these locations on the west coast are associated with breeding activities, mating and calving, but the humpback whale feeding grounds are in the Southern Ocean in the Antarctic.

Now, with *Whale Song*, we could venture down south in our fine ice-class steel-hulled expedition ship.

Our burning question with this Antarctic research was to determine the factors influencing the poor recovery of blue whales, particularly Antarctic blue whales, since the moratorium on whaling. We aimed to compare the feeding strategies of humpback whales, a species that had recovered well post-whaling, with Antarctic blue whales and record any major differences between the species on the Antarctic feeding ground. Are humpback whales opportunists? Are Antarctic blue whales fussy eaters? Antarctic blue whales have been protected from whaling for even longer than humpbacks, yet their populations have not recovered. How are differences in their feeding strategies affecting the success of Antarctic blue whales in terms of population? If their feeding strategies were similar, then what other aspects of their life cycle are threatening their survival? We hoped to begin addressing some of these questions with this foray into the icy Southern Ocean.

The research we had done on blue whales in the Perth Canyon years before with our colleagues John Bannister, Rob McCauley, Chandra Salgado Kent and Chris Burton had sparked our desire to know more. Rob McCauley had reported that acoustic recordings show that Antarctic blue whales, which spend the Southern Hemisphere summer in the Antarctic, also occasionally visit the Perth Canyon in winter, noting a distant chorus of their calls. The Perth Canyon continues to produce surprises!

During whaling, the population of Antarctic blue whales was reduced from some 300 000 to only around 6000 animals. Pygmy blue whales, usually found north of latitude 40 degrees south, fell from around 13 000 to around 6000. By comparing the feeding strategies of the abundant humpback whales with the more scarce Antarctic blue whales, we hoped we could get a handle on at least one of the issues preventing their increase.

Associate Professor Rob McCauley, our colleague in many cetacean research projects since 1995 and a world authority on underwater acoustics, was tasked with collecting the acoustic data and the krill biomass streams. Dr Russ Andrews, who we had met on the Great Barrier Reef earlier in 2013, brought precious satellite tags in his luggage. He had finely tuned the LIMPETs (Low Impact Minimal Profile Electronic Tags) to understand the movements of marine mammals without harming them. This had been Russ's focus for a couple of years, as he turned his biologist's hands to the tasks of a technician.

o

On 30 December 2013 *Whale Song* left the protection of Hobart's Prince of Wales Marina in the upper reaches of the Derwent with its experienced crew and seasoned researchers plus 34 baguettes, 34 chocolate bars, 50 kilos of meat, 5 kilos of fish, 710 coffee pods, 30 cos lettuces and 30 sliced loaves of wholegrain and ploughman's white bread, and many rounds of local cheese, together with 20 kilos of pinkeye potatoes and 40 kilos of rice ... I hoped we'd have enough to get us through until lunch!

We had christened this our WAVES (*Whale Song* Voyage for Ecosystem Studies) expedition. Blue whales and humpback whales would be our focus, and we aimed to document the difference in their feeding strategies using

BUILDING BLOCKS OF THE ANTARCTIC ECOSYSTEM

The rich ecosystem in the Southern Ocean is kick-started with cold water full of nutrients. As you would expect, in the summer months the water temperature is warmer than in the winter, but it is still rather cold. In fact, Simon Kenion, our second mate, had us wear our offshore SOLAS (Safety Of Life At Sea) life jackets religiously whenever we were on deck; as the sea surface temperature hovered around zero, a person accidentally flung into the ocean would only last two to three minutes, depending on their body weight. We would try our darnedest to stay out of the water but at the same time I liked thinking about the power of this amazing water. This is a thing I have noticed over half a lifetime of boating: the sea catches you in its intriguing embrace while also demanding a healthy respect.

The cool, nutrient-rich waters surrounding the Antarctic continent provide increased levels of dissolved oxygen. Together with the upwelling and the mixing of nutrients by currents (such as the Antarctic Circumpolar Current, the world's fastest flowing current, which travels eastward at 153 million cubic metres per second), a multitude of processes are initiated. During summer there is 24-hour sunlight. In short, it is the perfect environment for photosynthesis to occur in the microscopic algae (phytoplankton), which are present in the upper layers of the sea. This algal bloom provides nutrition for small animals, zooplankton or krill, which in turn commence the Antarctic food chain. A plethora of wildlife — seabirds, seals, porpoises, dolphins and whales — rely on summer krill patches, the basis of the Antarctic ecosystem.

a BioSonics echosounder to record the details of the krill patches they fed on. We had satellite tags to follow the whales' movements within and between krill masses, and would collect skin biopsy and photo-ID to identify individual whales.

As we left the Derwent, spitting rain tried to dampen our spirits, but it could not. Into the sunset we sailed, with pink-toned cumulus and heavily rain-laden cumulonimbus clouds above us.

I kept a daily blog throughout the voyage, and that evening I wrote:

> Curt and I dedicate this voyage to the late Grant Wilson, the previous owner of *Whale Song*. Grant's forward vision made this expedition possible and also the previous three years of science and adventures. We also acknowledge the involvement of John and Lyn Lally and David and Michelle Davenport from the outset of our career in Western Australia, as well as our wonderfully understanding families for supporting our crazy dreams ... Having envisaged and created our niche with Defence, listening to the Southern Ocean will be amazing, thanks Daffy.
>
> This expedition is an Explorers Club [EC] Flag Expedition, our WAVES Expedition is carrying Flag 69. This flag, which was first carried on an EC Expedition in 1937, will accompany us among the whales in the Antarctic. The 202 flags of the Explorers Club have been to the moon, the depths of the ocean and to

the heights of the mountains. It is with great honour and privilege that Curt and I carry this flag with this expedition and are humbly reminded of the courage and determination of the explorers on the previous five expeditions on which this flag has been carried.

Today we are leaving for the Antarctic. I am so excited I fear my soul will burst.

KRILL SOUP AND TAG SUCCESS

It was Day 16 on our WAVES journey, 14 January 2014. At the end of my night watch, I felt as though my eyes were dinner plates and I was totally wired with adrenalin. Being fully alert was all part of the process of our two-person watches, making sure we kept *Whale Song* safe. All was OK as we continued westward, ever hopeful of spotting whales. After a wonderful four-hour sleep, my second sleep of the night after my watch, I noticed the glassy calm sea and the bright sunlight. Only a few patches of fog came and went — yippee!

Curt popped his head in the door saying 'You'll want to see this ...' I dressed quickly and went on deck. A gorgeous iceberg like the Matterhorn towered ahead of us. This was a great photo opportunity. The whole crew — Curt, Dale, Simon, Rob, Russ, Sam, Resty, Inday, Tas and I — came out on deck, grinning and taking selfies. With immense pride, Curt, Tas and I held aloft the Explorers Club Flag 69 in the

Antarctic, the feeding grounds of humpback whales. Inday Ford, our research assistant, took photos and video of our exploring family with this beautiful flag. This really was a special moment as with deep respect we remembered and honoured the explorers of the past.

O

As we proceeded westward, the sea became a glassy grey — just wonderful. Fog rolled in and out, hampering our visibility, but almost as soon as it cleared, the call 'We've got a whale!' went down to the gang below in the main saloon. Very quickly all hands were on deck carrying all the fixings for our whale research — warm clothes, cameras, videos, satellite tag gear and biopsy equipment. The three blows of each surfacing were from a dark dorsal and we were looking at an Antarctic minke whale. Photo-ID was collected and we continued on.

In the clear visibility and even sunshine among the icebergs, we we approached a pod of possibly four humpback whales. Early on, as we followed them for satellite tagging, biopsy and photo-ID, we determined there were only three in the pod. Easily and effectively the three whales were identified by by their individual features as Captain Hook, Scar Face and Whitey. Naturally, the whale with a very hooked dorsal fin was given the handle Captain Hook; the animal with white crosshatch scars all over was named Scar Face; and the animal with a white patch behind its dorsal fin became Whitey. They were very interested in the abundant krill in the water, which showed as a huge band on the BioSonics computer screen. They rolled, surged and generally rotated every part of their bodies in elastic and fluid movements. It was wonderful to see a 15-metre whale weighing 40 000 kilograms target its food so carefully. Readying to carefully photograph the tag deployment with my 400-millimetre lens, I was delighted when their mobile bodies filled the camera frame. Right there were inflated heads, baleen plates hanging out of their mouths, their blowholes, 'arm-pits', ventral pleats, lip grooves, dorsal fins and twisting, rolling dives — I loved every bit of the whales.

In the fog and misty grey, the whales surged beneath the bow of *Whale Song*. Russ Andrews was standing in the bow basket clutching a Dan-Inject rifle loaded with a satellite tag secured to an arrow. Crouching behind him on the stepped platform of the bow basket, I was ready to photograph every aspect of the deployment. Fortunately, we were extremely well attired and comfortably toasty warm. Outfitted in hi-tech Eddie Bauer polar clothing, my core was really warm and, unbelievably, almost hot. With the sun breaking through the mist, the only reminder of our southern locale was the cooling of my fingertips, simply because I could only wear one pair of gloves to use our equipment. Keeping your hands dry is a key factor for comfort and wellbeing in watery polar activities. My purple rubber gloves over the inner thinsulate pair worked a treat, but looked a little odd. I was sure the whales weren't bothered by my funny hand-wear!

The beautiful pink and black-and-white expandable throat grooves on show before us were the splotched individual patterns of the whale's belly, easily visible when the whale was not feeding. The delicately pink-toned lines were the longitudinally oriented, expandable folds or pleats between the outer black-and-white belly pigmentation. Together they extended almost two-thirds of the animal's body length. The pleats and the expansion they afforded allowed a humpback whale, like other baleen whales known as batch feeders, to scoop huge volumes of seawater into the lower jaw and filter their tiny prey — in this case, krill (*Euphausia superba)*, a small shrimp-like crustacean about the size of a single finger joint that flourishes in the Antarctic. After this humpback whale carefully negotiated its way through a dense krill patch, the process of filtering the krill through its baleen plates began. Once the lower jaw was expanded and full of water and krill, it was brought back to meet the upper jaw with the filtering baleen plates and the fringed inside edge of the plates. Water poured from the mouth, the 2-tonne tongue moved backwards and forwards in a series of muscular movements, the krill gathered into a food bolus and swallowed down the grapefruit-sized oesophagus. Krill that spilt from either corner of the lead whale's mouth was engulfed by the two whales following closely behind it in an echelon formation. This was cooperative feeding at a new level.

Captain Hook rolled under *Whale Song*'s bow into perfect position. With a squished bang the tag was deployed and a biopsy collected. The deck crew scurried to keep an eye on the arrow and the biopsy sample in order to collect this floating hardware. I made notes and got photos of Captain Hook with its new gear. We repositioned and once another tag was readied we closed with the pod again. Making sure we got a new whale, I called names of the whales as they surfaced and Russ relayed them by headset to Curt, who was driving from the Portuguese bridge.

'The one we want is Scar Face, Captain Hook is on the right — here it is!'

'Are you ready, Mich?' Russ asked, and quick as a wink the second tag was deployed on Scar Face and a biopsy collected too. Yahoo!

We followed the pod until 11.30 p.m. as the whales continued to fluke-up dive side by side and then surface together a minute or two later, side-lunging through the krill with pectoral fins, inflated heads, open mouths and water pouring out of their mouths right next to each other at the surface. A sight to behold!

Rob noted that there must be an advantage for the pods to be so closely oriented. By diving together and surfacing in such a coordinated and synchronised manner, they must be concentrating a patch of krill so that when they're surfacing they've trapped the krill against the natural barrier of the water's surface. Curt pointed out that it was not just coordinated, it was cooperative and the echelon technique could be a key to their success, similar to the way humpback whales cooperatively use bubble nets while feeding in the Arctic. Through the surface they lunged, taking huge mouthfuls of food. I quite liked the idea that they were working together here in the 'kitchen'. After all, eating is a nice social activity.

The behaviour of Captain Hook and Scar Face was very repetitive. (The third whale in the pod we had originally sighted, Whitey, had affiliated with a pod nearby.) They surfaced together, lunge-feeding through the krill, then three to four blows later they fluke-up dived and then repeated the sequence over and over again as they filled their bellies with krill — the keystone of the Antarctic food chain. The BioSonics echosounder showed the krill as a dense band stretching from the surface to 40 to 70 metres deep. The combination of the real-time dive profiles of the whales with the real-time echosoundings of the krill biomass they swam through created an incredible dataset.

The two whales surfaced together and continued to feed in their incredible synchronised style well beyond my 9 p.m. bedtime. We followed, collecting krill density data to analyse with the tag tracks. Knowing we had whales feeding right near the bow was part torture and part delight! Wanting to be on deck with them was the torture, but I needed to rest, and the delight was sleeping knowing that Captain Hook and Scar Face were feeding nearby.

ANTARCTIC FACTS

Large icebergs 10 kilometres or more long are given code names by the US National Ice Center that relate to the quadrant and consecutive sighting. For example, an iceberg called C-16 could have been the sixteenth iceberg encountered in 'C' zone, 180 degrees to 090 degrees east (i.e. in the western Ross Sea/ Wilkes Land).

O

Pure ice is blue because it contains no air bubbles, which reflect light, and therefore it efficiently absorbs the other wavelengths of light, particularly red light. When sunlight, which is white, shines upon pure ice, only the blue light is reflected back to your eye. The more bubbles the ice contains, the more white it appears; the fewer bubbles, the bluer it appears.

O

Casey Station, one of three Australian Antarctic Stations, was established in 1969, and is located in a beautiful area called the Windmill Islands. It accommodates 88 people in summer and 19 in winter.

O

'Big eye' is Antarctic slang for a period of sleeplessness, most often caused by the 24-hour daylight of the Antarctic summers, but it is also triggered by the endless darkness of winter.

TRACKING TINY

The bergs were beautiful and we got to know each of them as we spent the day searching for a pod of whales that was feeding. As we headed towards a fantastically sculpted 'chair', a designer seat upholstered in snow, a humpback whale began breaching. We followed, and as we approached the whale it surfaced next to us. It appeared a tad confused — I think it had been trying to attract a pod of three adults travelling nearby but had attracted us first! Soon the three whales joined Barnacle Bill (the Breacher) and these we named Nick, Flat Top and Tiny. Tiny was tagged with a Whale Lander tag, which records precise heading, yaw, pitch and roll data, and was thus given the handle Tiny Lander. This tag is designed to come off and we had to retrieve it to gather the data via cable download. With this in mind, we followed Tiny Lander and its cronies. That afternoon and into the evening, a whale soap opera unfolded. Tiny Lander joined Hi Point and then pretty soon we had 12 different whales all swirling around in the vicinity — including Barnacle Bill, Pointer and

Chunk — affiliating and disaffiliating with fluidity. Several more pods joined and we noted we had Whitey and Diatom, a smaller whale covered in yellow diatoms on its belly and tail flukes, very close and engaged in much social contact. Perhaps Diatom was a calf of last year, which would explain its rambunctious behaviour. We followed behind keeping a record photographically, and Inday also videoed all the activity.

All through the dusky pink light of the Antarctic 'night' we followed Tiny Lander, who travelled alternately with White Racer and Barnacle Bill or Lionel and Square. Tiny Lander's tail fluke was very distinctive with an easily recognisable black 'butterfly' or # 55 fluke pattern on the underside. By its small dorsal fin and the distinctive markings of the tail with each fluke-up dive every few minutes, we knew we were following the correct whale. The watch teams had to concentrate to stay with Tiny Lander because several other humpback whale pods were scattered around the icebergs and all were slowly mobile. The longer the tag stayed attached the more interesting the data became, and the more important it was for us not to lose it.

In the morning Curt woke me an hour earlier than usual.

'You have to come and look on deck, we need to find this tag.'

Becoming fireman-like in my fast dressing technique, I raced on deck as quickly as possible given the five layers required on my torso and three pairs of pants plus ice boots, life-jacket, binoculars and camera. *Right, I'm awake. What's up?*

The Whale Lander tag was designed to come off after several hours for data retrieval and indeed it had detached from Tiny Lander at 4.45 a.m. The problem was that Tiny Lander had given the 3 a.m. to 6.a.m. watch crew the slip, and *Whale Song* was no longer following him … Uh oh! Russ checked the specialised hardware and directed us to the last known position. As we motored around in the dull Antarctic near-twilight I checked with Russ exactly what we were looking for. The Whale Lander tag was the same diameter as an orange, four centimetres deep and putty-coloured except for nine or 10 reflective patches on the flat top and around the sides. It would be floating right at water level with no profile showing. Wow, that is going to be hard!

Curt drove *Whale Song* in a classic search pattern around the last known position. Peering into the sea, I concentrated really hard, straining my eyes to find this precious package. I was delighted when it came into view.

'There it is! It's here, at two o'clock, 50 metres! Wahoo!'

Immediately Curt drove *Whale Song* towards the tag and Sam retrieved it from the icy waters of the Antarctic with our pool net. It was loaded with 13 hours of heading, yaw, pitch and roll data from Tiny Lander, a feeding humpback whale! This was the first application of a Whale Lander tag on a humpback whale and the first Whale Lander tag in the Antarctic.

HUNGRY WHALES AND FUSSY EATERS

We found many humpback whales in the Antarctic. Standing on the foredeck with a 360-degree vista of a flat platinum sea and silvery sky broken by icebergs of every size and shape, not to mention the steamy blows of at least 13 animals in four pods, I kept pinching myself. Am I really here? Naturally, the science was paramount and we received 12 streams of data continuously, building a huge database of the movements of the feeding humpback whales in relation to the abundant krill. But it was impossible to separate the thrill and euphoria of gathering the science from the beautiful fluid movements of the synchronised whales in such a stunning setting. Being passionate about our work every day had got us this far. This work was no exception.

The behaviour of the feeding humpback whales was absolutely fascinating. We were able to document the different types of behaviours, which we designated by whether they occurred on feeding or breeding grounds, and whether they were surface active or surface passive. The surface passive behaviour and the surface active behaviour humpback whales displayed on the feeding grounds were very different compared with those we regularly recorded on the breeding grounds. Following whales day and night, we made unique behavioural observations. During the morning and into the early afternoon the whales exhibited surface passive displays. These were pedestrian behaviours by nature. The whales surfaced through the water, exhaled/inhaled, dived shallowly or with fluke-up dives, just moving slowly, nothing flash or showy. All their movements were purely in a dorsal-ventral orientation (up/down) without any splashy displays of activity. They were low key and resting, just passing time until they commenced their next round of feeding activity. For approximately 12 hours from around 3 p.m. until the wee hours of the morning, the whales shifted gear into feeding mode. Beneath the bright Antarctic almost 'twilight' skies, this behaviour was spectacular and extremely interesting.

The surface active behaviour exhibited by feeding humpback whales on the feeding grounds involved a totally different set of behavioural displays. When feeding humpback whales displayed surface active behaviours their body orientation changed from purely dorsal-ventral movements to incorporate twists, turns and full rolls and rotations under or at the surface. These highly mobile displays enabled the whales to successfully navigate into the krill patches and maximise the krill captured when lunge-feeding. Breeding ground surface active behaviour involved aerial behaviours such as charismatic breaches, fluke slaps and pectoral fin slaps, which are noisy and percussive and associated with competition and communication. The active aspects of the whales' behaviour on the feeding ground were totally different, as they were all about directing the whales' attention to their prey, rather than competing with each other. These feeding whales were in stealth mode: hundreds of 40 000-kilogram humpback whales were quietly sneaking up on the krill. Imagine that, if you will! With careful and clever twists and turns they rotated their bodies remarkably well to surface successfully with mouthful after mouthful of tonnes of krill. The only sounds during these feeding ground surface active behaviour displays of rotating and twisting and turning were the gentle surges of water on the table-top sea as the bodies rolled and twisted around the krill balls.

A wonderful example of the nuances of humpback whale behaviour in different locations was a boisterous little humpback whale calf. A calf of that season, being four to five months old, it breached over and over relentlessly. It photo-bombed Tiny Lander, our whale with a Whale Lander tag, who we followed for the full 13 hours from attachment, and it even breached mischievously right near a large flock of resting light-mantled sooty albatross. This calf was not being quiet and it was certainly not being quietly 'active' like the adults. It seemed it had not yet learned the subtler aspects of being a feeding humpback whale. We noted that its mum politely ignored its breaches of protocol, but no doubt in time this calf would learn to conform.

The upshot of the data we collected on WAVES was that, as we suspected, humpback whales feed differently from blue whales. In pods of multiple individuals, humpback whales use an echelon feeding technique, cooperatively increasing their chances of gathering the available krill. They precisely targeted large krill patches, assisting each other by using sensory information to gauge the surrounding water temperature. The lumps and bumps called sensory tubercles that are scattered across the heads of humpback whales provide water temperature and water quality information. From our studies in the Antarctic that season, this variable water temperature was the key factor in determining where these humpback whales fed. The cooler the water, the more whales were observed and the more they fed, swirling through the dense krill patches. Their long, oarlike pectoral fins enabled them to rotate, twist and turn quickly and thus carefully maximise their capture of krill.

By contrast, blue whales feed singularly and opportunistically, even apparently 'choosing' whether to expend energy in the process of gaining energy in relation to the size of krill mass presented.

If it turns out that blue whales are fussy eaters, then sadly, this could be their downfall if the world's oceans don't have the broad concentrations of krill they once had. Humpback whales defecate as they feed, fertilising the rich algal zones near the melting ice-edge that in turn creates the food krill graze upon. A concentrated, self-perpetuating feeding ground results. If blue whales need a similar cycle of self-fertilised feeding areas, but on a global scale, it could take hundreds of years to wind back the damage that whaling has done — and simply may not be possible if we humans continue to damage the planet.

SPERM WHALES

'Hey Mum, I've got a blow. It's left-angled — do you think it's a sperm whale?'

Tas's excited voice came over the UHF radio from the flybridge, where she was on her hourly Cetacean Observation Watch. In the wheelhouse on the level below, Curt immediately steered *Whale Song* in the direction of her sighting. Sure enough, with the next surfacing we clearly saw a lone male sperm whale. This bull was a huge animal, perhaps 18 metres long.

'Twenty-seven, 28, 29, here are the flukes!'

We were counting the blows following an age-old tradition. The whalers determined that the number of blows displayed at the surface directly indicated the number of minutes that a sperm whale would stay down. Exactly 29 minutes later this old bull surfaced again, perfectly on cue. Sperm whales fluke-up on almost all deep dives after travelling at the surface, and this bull was no exception. Its broad, near triangular flukes with rounded tips and straight trailing edges were lifted high and almost vertically out of the water as it dived, perfectly framed between bergs.

Sperm whales, the largest of the odontocetes or toothed whales, are fascinating in their morphology and their intricate social structure. As their classification suggests, toothed whales are characterised by their cone-shaped teeth. They also have a single blowhole and utilise sonar for prey detection and foraging. Researchers Hal Whitehead and Luke Rendell have dedicated their lives to unravelling the mysteries of these beautiful animals, including detailing their extraordinary society. Female sperm whales gather in nursery pods of mothers and calves in the tropics, often numbering from 20 to 30 or even as many as 50 individuals. Sexually mature but as yet non-breeding males form bachelor pods after leaving their natal groups. At age 28 these bachelor males separate and roam alone across ocean basins, associating with the highly social pods of females for only brief periods of time. Being polygynous, the adult males employ a 'roving' strategy for mating. Female sperm whales reach up to 12.5 metres in length compared with the larger males, who reach up to 19.2 metres. Only the large adult male sperm whales migrate to the polar regions to feed. Sperm whales are the most sexually dimorphic of cetaceans, the males weighing almost three times as much as the females.

The whale Tas had found swimming among these undeniably beautiful icebergs was an old lone male. Scarring across its huge head and body, its dorsal fin set well back on the dark brown wrinkle-patterned skin and the lack of white or yellow-toned calluses on the dorsal hump indicated the age, class and gender of this animal. The initial sighting cue Tas had seen, a left-oriented blow as seen from behind, was a clear clue to its species. Sperm whales possess a single S-shaped blowhole that is slightly offset to the left. In light winds the left-angled blow is totally distinctive. Sperm whale on the left, icebergs on the right — where to look?

WHERE HAVE ALL THE BLUE WHALES GONE?

Each day we saw sperm whales, minke whales and more humpback whales, a veritable mix. With every passing day, 'Where are the blue whales?' was the question on everyone's lips. Acoustically we were monitoring four times a day with sonobuoys, *really* hoping for some blue whale detections. However, we could only hear lots of sperm whale foraging calls, seals shrieking and icebergs cracking. Were humpback whales dominating these areas? And where were the three-quarters of a million Antarctic minke whales?

As we followed the soft ice-edge westward through East Antarctica, the most remote area in Antarctica, we began to hear some blue whale calls. Had we finally found a hotspot for these whales? We were relieved and thrilled all at once. As we approached one particular bay, which we excitedly dubbed 'The Bay of Whales', Curt received some urgent advice from the duty forecasters at the Australian Bureau of Meteorology (BOM).

Being a Volunteer Observer Ship, we were providing weather observations by satellite email four times daily directly to the BOM. We pumped the weather observations in at our location assisting the model makers; it was a weather circle about the weather cycle. So while we were down south, each day Curt provided the BOM Antarctic Forecasters with three possible positions for *Whale Song* at noon in 24 hours' time. Curt would then receive weather forecasts for those locations from Tina Donaldson and Bill Plant at the BOM via satellite email, which allowed Curt as the master of our vessel to make safe course choices in relation to the upcoming weather conditions. The information from the BOM was pure facts, no advice was being given. Curt made his own decisions based on the three options regarding the next day's strategy and speed.

Usually the daily BOM inbound communications were concise emails regarding the weather approaching from the west for the next four days, with an accompanying series of *Whale Song*–centric pressure and wind maps. A second and equally valuable resource was provided by Neil Young, an Antarctic ice expert based in Hobart, who sent daily ice movement data via infrared and real colour satellite maps. Usually Neil wrote no more than two lines of text to accompany the maps. When Curt mentioned to Neil that we could hear blue whales and were keen to venture inside a bay formed by an iceshelf to the south-west, he was startled to receive a *one-page* commentary in return, including the helpful instruction: *Under no circumstances should you enter that bay, there have only been two ships in history that have gone into that bay and both have been trapped by ice. You may not return.*

Because ice moves in a westerly direction with the prevailing winds inside an east-facing bay (the orientation of our 'Bay of Whales'), our vessel would surely become stuck in the drifting ice, just as two other vessels had done in the 1960s and 1980s with dire consequences. The 2013–2014 summer season had already proved to be a bad season for ships, with the *Aurora Australis* and *Akademik Shokalskiy* getting stuck in the ice. Curt, our careful ship's master, was grateful to have Neil's expertise so readily at hand.

We dutifully renamed this bay 'The Bay of Death' and that afternoon made our way to the north-west, away from the continuous soft ice-edge where humpback whale pods swam tantalisingly among the small bergie bits, and *away* from the few blue whale calls we had detected. It was excruciating to pass by a bay that may have held the answers to so many of our questions. It was heartbreaking for Curt and all of the crew. In the end, however, the safety of the expedition was paramount and it was the right decision.

As we travelled homeward, concentrating on dodging nine low-pressure weather systems (in addition to the four we had danced around on the transit south), we were pleasantly surprised to find a pygmy blue whale and an Antarctic blue whale in the Sub-Tropical Convergence Zone. What were they doing here at the end of January? Did they regularly feed here and then go into the Southern Ocean in February and March? Like most scientific investigations, we were returning with more questions than answers — especially regarding blue whales.

A STORMY HOMEWARD PASSAGE

'What has happened? You've turned north. Why are you coming home?'

Micah was on the satellite phone, her voice slightly panicked. We had made a turn to starboard ten minutes earlier and indeed we were heading home. The change of course had been instantly visible on our online ship tracker that Micah was watching from Perth. Curt had logged on that morning and the usual email from the BOM forecasters had provided worrying news: a low-pressure system they were calling 'The Beast' was headed directly our way. Should we hide for five or so days at the soft ice-edge to let 'The Beast' pass?

Curt felt discretion was the better part of valour, and so at 9.00 a.m. on 16 January 2014, we turned 90 degrees to starboard and sailed north towards Fremantle. For the previous 15 hours we had hidden behind a 1.5-nautical-mile-long tabular iceberg that looked like a meringue. As the wind and sea buffeted it, multitudes of growlers and bergie bits broke off, forming a floating ice fence. Curt stayed at the wheel for the full duration of this 60-knot blow as sleet, snow and icebergs surged around *Whale Song*. As he tried to stay as close to the berg as possible to get some protection from the wind, the icy debris became another hazard. Bergs the size of our boat came at us, expertly dodged by Curt's careful hand at the wheel, but the hundreds of smaller chunks of ice were unavoidable and clanged, scratched and scraped against the hull. The palm-sized Spot Tracker unit that Russ had attached to the railing on the flybridge sent a latitude and longitude message every ten minutes to registered receivers, such as Russ's and our families. This was an effective safety net that Micah monitored continuously.

We were homeward bound, a journey that entailed dodging no less than nine low-pressure systems. Some had been given endearing names by our BOM friends such as 'The Son of the Beast' and even 'The Mother of the Beast'. These names had Curt pushing our little vessel as far from their paths as possible. There was much hooting and hollering when we arrived at the Fremantle Fishing Boat Harbour to welcome us home. The relief of returning safe and sound was palpable.

During 31 days at sea, our WAVES expedition travelled 5518 nautical miles south from Hobart and westward along the soft ice-edge and then north to Fremantle. Transiting through temperate and polar seas we observed 12 species of cetaceans, identified 30 species of seabirds, deployed six satellite tags on humpback whales, collected nine humpback whale biopsy samples, submitted 120 BOM weather reports, recorded biomass data for nine days straight, launched 140 sonobuoys and heard sperm whales clicking, seals shrieking and icebergs calving. And I took 30 000 photographs!

The wonder, spectacle and power of this beautiful ecosystem all starts with cool water. It's the Antarctic magic.

WEIRD WHALE FACTS

The grey whale is the state animal of California. Gray whales are very friendly and like to be rubbed with brushes and brooms.

o

The lungs of a humpback whale are the size of a VW Kombi. When whales strand and lie on their bellies on the beach, sadly their lungs become pierced by their ribs. In the ocean the water supports their bodies as they swim.

o

Long-finned pilot whales strand most commonly, often in the hundreds. There are many reasons why whales strand. One widely accepted theory is that when a family member is in trouble, the entire pod will strand together as they have very strong social bonds.

o

A humpback whale is as long as a bus, 16 to 17 metres. A humpback calf is five metres at birth, the size of a Toyota Camry.

o

Belugas, the white whales, sing songs and were once called the 'sea canaries'.

o

The 2-metre spiralling tusk of the Arctic narwhal is really a modified tooth. Only males possess this tusk, which is often used in competition for females.

O

The humpback whale, which is a baleen whale or a 'moustached' whale, has teeth as a foetus. These foetal teeth, 28 in the upper jaw and 42 in the lower jaw, are reabsorbed and disappear before birth.

O

When a humpback whale blows, it exhales 90 per cent of its lung capacity at 480 kilometres per hour. Humans only exchange about 13 per cent of their lung capacity with each breath.

O

There are approximately 750 000 minke whales in the Antarctic. When many of the large whale stocks (blue, humpback, fin, sei and right whales) were almost wiped out by whaling, the minke whale populations increased due to the abundance of their food supply — krill. Minke whales were not of interest to whalers due to their small size. Unfortunately, now their abundance has the whalers interested ...

O

Young male sperm whales travel in large loose-knit bachelor pods from ages 6 to 27. The large older males prefer a solitary existence. Sperm whales can dive to a depth of 3 kilometres and remain submerged for up to two hours.

o

The oesophagus of a humpback whale is only about the size of a large grapefruit. A dead whale in Alaska was found to have seabirds lodged in its oesophagus. The birds are attracted to the feeding frenzy by the panicked fish. The whales create a bubble-net by exhaling and swimming in a wide spiral towards the surface, engulfing the huge feed-ball of herring, krill and capelin.

o

The killer whale isn't really a whale but is the largest of the dolphins. Killer whales, or orca, hunt other marine mammals such as Dall's porpoises and seals. Records have also shown deer remains in the stomach contents of dead orca.

o

Southern right whales were so named because they were the 'right' whales to hunt. They are slow, lumbering swimmers that the whalers could easily catch. By 1900 right whales were considered virtually extinct. In 2005 the population of southern right whales off Western Australia and South Australia was recorded as 2100 individuals. This population is believed to be approximately 17 per cent of the global population.

o

The baleen of a bowhead whale is the longest of all species, reaching over 3 metres. Often the baleen plates show a green iridescence at night.

o

Western Australian humpback whales migrate 6500 kilometres between their 'bedroom' in the Kimberley and their 'kitchen' in the Antarctic. Their 'corridors' often pass through international oceans.

o

Southern right whales have unique patterns of callosities on the top and sides of their heads that are as individual as our fingerprints.

o

The fin whale has asymmetric pigmentation on the right side of its head. The white colouration includes the lower 'lip', mouth cavity and some of the baleen plates.

o

To estimate the size of large dolphin groups, scientists count the number of dolphins seen at the surface, then multiply by three.

POSTSCRIPT

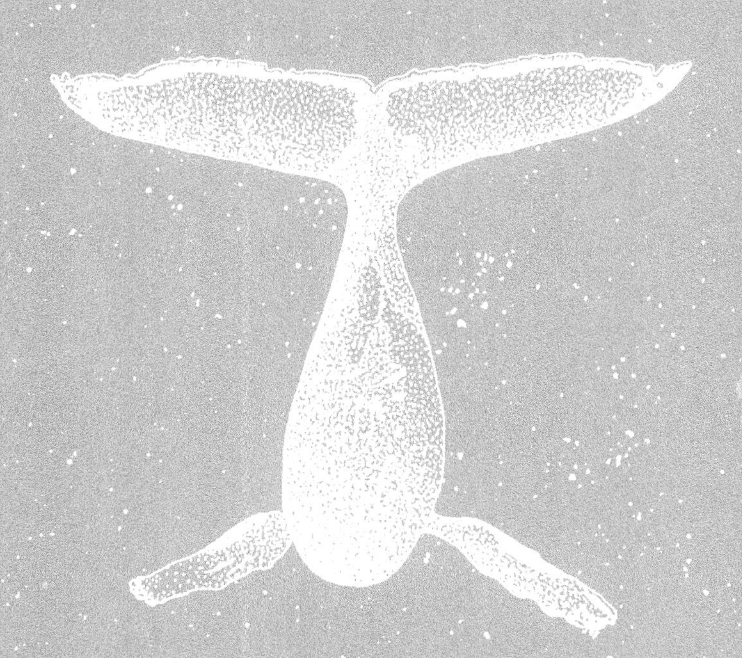

TOO MUCH FUN IN THE SUN

I have loved every minute of our sails up and down the coast in all of our vessels. Somehow my selective memory blocks out the details of the very rough stuff and I focus on the lovely flat days, those days that cetacean biologists live for. But I also just love being at sea.

It hasn't all been beer and skittles, however. In the 1970s and 1980s, children in New Zealand (where I grew up with my mum and my brother, Ross, after we moved from my birthplace, Sydney), were encouraged to play outside in the 'sort-of' warm sunshine. I needed no encouragement. Barely able to sit still, I was an active and on-the-go child, walking Penny, our golden retriever, twice a day. Cycling and swimming at one or all of the five beautiful eastern suburbs beaches was just part of a normal summer's day. You can guess where I am going with this: no one discussed skin cancer then and we were unaware of the hole in the ozone layer that was perfectly positioned right over New Zealand. My fair skin bronzed easily in the long summer days — but more recently my mother's Scottish ancestry has trumped my father's Belgian heritage.

Just after my mother's passing in October 2005, several of my freckles and moles began to change at an alarming rate. Major differences in shape and density occurred within just three weeks. My personal theory is that with the increased stress of my mother's death, my immune system, already compromised by unwitting childhood sunburns compounded with unwise adult exposure, couldn't fight any more.

Surgery is the solution for *in situ* skin cancers, those naeve or moles that have not metastasised into other regions of the body. By July 2006 I was able to go swimming again after six months of wearing bandages covering up to 50 stitches at three sites. Two moles were melanomas (Level 1, the lowest of the four levels) and thus I had double surgeries at those two sites to make sure there was at least a 10-millimetre margin of clear tissue around the offending spots.

I have ongoing three-monthly examinations and six-monthly photo-mapping, which uses computer imagery to match the spots and detect changes. I totally appreciate the care taken by my doctor, Dr Lester Cowell, and his technical surgery photographer, Monika Cowell. It's 11 years since my first surgery in January 2006 and I have now had two dozen procedures. There is no doubt that thus far these procedures have kept me going to gainfully and joyfully annoy my family, friends and fellow beach-walkers.

I am extremely grateful for these last 11 years. When I was first diagnosed I had no idea how much time I had left, if I had two weeks, two years ... For three days I was inconsolable and deeply worried about the path ahead. Then on the fourth day I said to myself, 'That's enough crying, it's time to get on with living.'

Now I hear myself echoing my mother's words to my own daughters: 'Cover up, put on sunblock, be careful.' I did wear hats and sunblock during our Pilbara and Kimberley fieldwork, but most likely the damage had already been done. I have always loved a good tan, so all of this is self-inflicted. Now I live by the creed: 'Don't mess with the sun or it will mess with you.'

Many good quality UVA and UVB blocking sunscreens and even fake tans are available now, as well as stylish rashies for swimming and surfing. Given the greater awareness of the dangers and countless campaigns to cover up in the sun, there is a possibility for this generation to be safe. But it might be more complicated than we realise. There is also a possibility that our liberal use of sunblock at the beach and before swimming in coastal systems might be harming the very places we enjoy. Physical barriers such as hats and clothing could be better for the health of humans and for coastal ecosystems. Don't be complacent, cover up. If just one person can avoid the scars and discomfort of surgery, this rant is worth it.

OF SEA

Oh the sea, you angry sea
What is your mood?
What is your game?
This strong current — Agulhas current, so free

Not ours to tame
Taking us on a southern course
To southern seas and iceberg clime
Like riding a wild, bucking horse
We shall reach Mauritius — in due course...

We ride, we buck, we glide and slide
On hills of blue and black and grey
Up and o'er white mountain caps
And peaks of turquoise water ride
On and on 'til end of day
When birds' and sun's energy lapse

The sea, the endless sea
Of endurance, exploring let us be
At one with you all the way
But do not let us unduly pay
For passage on your bobbing waves
Above aeons of sailors' watery graves

I wish to sing a song to you
And calm you to a gorgeous blue
With puffy clouds above the glass

We will hug and rest at last
In peace and calm seas across the oceans vast

Above rising seamounts and plunging canyons deep
My soul with you shall always keep
The oceanic dance — not just a glance
But a truly deep, ocean trance.

Thanks for the blue, green and gold
Never shall your views grow old
For upon the vasty seas we shall
Always have stories to tell

Of dolphins spotted, leaping and devout
Around us and each other about
Of whales, Bryde's, sperm and humpback
That with their flippers and flukes whack

The interface of sea and sky
In which they live and try to fly!
Since air within their lungs inhale
Brings them to the place we sail

We sit atop the water's edge
Where mammals dally, as on a ledge
To swap the old and grab the new
From lungs' depths to open air
A pillar of gas ... 'Hey, over there!
Got a blow, looks like two!'
Now tell me, what sort are you?

We catch a glimpse of a deep sea friend
Whose life on the surface will only spend
One tenth above, nine-tenths below
So, to the depths we must surely go

To truly understand their passage
Of life — and read their message
That this sea is home to ancient creatures
But we just see the stock market features

When considering exploring for mineral finds
Oh — please take off the blinds
That dull our sight and goals that hinder
Which make us bring age old beauty to cinder

We thank the sea and all her moods
For all the glory she exudes
Give praise for sea and spray and ocean
And all the calm or rising motion
Thank you sea, and glory be.

Micheline Jenner
Written 18 Jan 2011 at 34 degrees 26.9 S and 29 degrees 03.7 E.

ENDNOTES

HUMPBACK WHALES

First breath

commonly sighted as they head for the Kimberley calving grounds over 1000 kilometres further north to give birth

Jenner, KCS, M-NM Jenner and KA McCabe (2001) 'Geographical and temporal movements of humpback whales in Western Australian waters', *APPEA Journal* 38(1): 692–707.

the fin, which is mostly cartilage, is folded over

Jefferson, TA, MA Webber, and RL Pitman (2015) *Marine Mammals of the World*, Academic Press, London, San Diego, Waltham, Oxford, p. 608.

this temperature makes significant difference to the energy requirements for newborns

Perrin, WK, B Wersig and JGM Thewissen (2009) *Encyclopedia of Marine Mammals*, Academic Press, p. 1352

Leather, H and J Leather (2016) *Cambridge Checkpoints VCE Biology Units 1 and 2* third edition, Cambridge University Press, p. 276.

Most mammals give birth at night

Feldhamer, GA, LC Drikamer, SH Vessey, JF Merritt and C Krajewski (2015) *Mammology: Adaptation, Diversity, Ecology*, JHU Press, p. 768.

Several years later, a research group in Madagascar documented the birth of a humpback whale calf

Faria, MA, J De Weert, F Pace and FX Mayer (2013) 'Observation of a humpback whale (*Megaptera novaeangliae*) birth in the coastal waters of Sainte-Marie, Madagascar', *Aquatic Mammals* 39(3): 295–305.

Female whales nearing their delivery time may well be sensitive to noise

NRC (VS) (2003) *Ocean Noise and Marine Mammals*, Committee on Potential Impacts of Ambient Noise in the Ocean on Marine Mammals, National Academies Press, Washington DC.

the Western Australian humpback whale population had steadily grown to between 33 000 and 36 000

Salgado Kent, CP, C Jenner, M Jenner, P Bouchet and E Rexstad (2012) 'Southern Hemisphere breeding stock D humpback whale population estimates from North West Cape, Western Australia', *Journal of Cetacean Research* 12: 29–38.

When whaling ceased in 1963, the population had been less than 500.

Chittleborough, RG (1965) 'Dynamics of two populations of humpback whale, *Megaptera novaeangliae* (Borowski)', *Australian Journal of Marine and Freshwater Research* 16: 33–128.

Bannister, JL (1994) 'Continued increase in humpback whales off Western Australia', *Report of the International Whaling Commission* 44, 309–310.

Using high-resolution cameras, readily available these days, three images are taken of each whale.

Katona, S, B Baxter, S Kraus, J Perkins and H Whitehead (1979) 'Identification of humpback whales by fluke photographs', Winn, HE and BL Olla (eds) *Behavior of Marine Animals*, Springer, Boston, Massachusetts.

we produced a humpback whale population estimate of between 2000 and 3000 individuals.

Jenner, KCS and M-NM Jenner (1994) 'A preliminary population estimate of the Group IV breeding stock of humpback whales off Western Australia', *Reports of the International Whaling Commission* 44: 303-307.

newborn humpback calf who would spend almost 30 per cent of each day suckling

Videsen, SKA, L Bejder, M Johnson and PT Madsen (2017) 'High suckling rates and acoustic crypsis of humpback whale neonates maximise potential for mother-calf energy transfer', *Functional Ecology* doi.10.1111/1365-2435.

Jenner and Jenner unpublished data.

when their adrenalin levels are lower and their serotonin levels are higher

Chan, S and M Debono (2010) 'Replication of cortisol circadian rhythm: new advances in hydrocortisone replacement therapy', *Ther Adv Endocrinol Metab* 1(3): 129–138.

have been observed, photographed and studied while gathered in these protected south coast bays

Burnell, SR, R Pirzl (2001) *Ecology and behaviour of southern right whales: Head of Bight, South Australia, Report*, p14.

A shaky whale nursery

a surface passive behaviour known as tail sailing or fluke extension

WDC Ethogram *Humpback Whale Behavior 2013* (pdf).

whalesenseblog.files.wordpress.com/2013/04/ethogram_mn.pdf.

Humpback whales are renowned for producing the most complex repetitive yet changing songs in the animal kingdom

Payne, RS and S McVay (1971) 'Songs of humpback whales', *Science* 173 (3997): 585–597.

males challenging other males in battles over females (considered dominance polygyny)

Darling, JD (2001) *Characteristic Behavior of Humpback Whales in Hawaiian Waters, Report for the Hawaiian Islands National Marine Sanctuary, Department of Land and Natural Resources*, p. 61.

A female might judge and be attracted to a suitor by the length of his song (known as lek polygyny)

Darling, JD (2001), p. 61.

they exhibit reverse sexual dimorphism, meaning that adult females are larger than

Clapham, PJ (1996) 'The social and reproductive biology of humpback whales: an ecological perspective', *Marine Mammal Review* (21)1: 27–49.

an adult accompanying a female humpback whale with a calf, usually referred to as an escort, is a male

Darling, JD (2001), p. 61.

WhaleSong (Before it's Too Late series) *award-winning documentary produced by Storyteller Productions, 1997.*

Storyteller Productions (1998) *Before It's Too Late 5: WhaleSong*, 56 mins.

Pec fin ride

they have the same bone structure as our arms (radius, ulna and humerus), our hands (carpal and metacarpal bones) and fingers (phalanges).

Watson, L (1981) *Sea Guide to Whales of the World*, Elsevier-Dutton, New York p. 302.

these raised protuberances have shown that they provide lift while the whale swims

Fish, FE, PW Weber, MM Murray and LE Howle (2011) 'The tubercles on humpback whales' flippers: application of bio-inspired technology', *Integrative and Comparative Biology* 51(1): 203–213.

The scientific name of humpback whales is Megaptera novaeangliae, *which translates from the Latin as 'big-winged New Englanders'.*

Watson, L (1981) p. 302.

the males jostling for position to be the female's primary escort

Parsons, ECM, AJ Wright and MA Gore (2008) 'The nature of humpback whale (*Megaptera novaeangliae*) song', *Journal of Marine Animals and their Ecology* 1(1): 21–30.

calving … is associated with the tropical Kimberley region during the months of August and September

Jenner, KCS and M-NM Jenner (1994) 'A preliminary population estimate of the Group IV breeding stock of humpback whales off Western Australia', *Reports of the International Whaling Commission* 44: 303–307.

It is documented that male lions will kill cubs that are not their own when mating with a female

Packer, C (2000) 'Infanticide is no fallacy', *American Anthropologist* 102(4): 829–831.

One crazy whale

the genital slit is located closer to the anus and is flanked by a pair of 4- to 5-centimetre mammary slits

Watson, L (1981) p. 302.

the tapetum lucidum, *also employed by cats in night hunting, maximises light by redirecting it back through the retina so that it is reflected twice*

Schwab, IR, K Cartlon and BS Yuen (2002) *Evolution of the Tapetum*, Trans Am Ophthalmol Soc (100): 187–200. (Schwab et al 2002)

her pupil was naturally shrinking to protect her eye from the bright rays of sunshine at the surface

http://www.whalesforever.com/whale-senses-sight.html.

it is believed whales see mostly in black and white rather than colour
Meredith, RW, J Gatesy, CA Emerling, VM York and MS Springer (2013) 'Rod monochromacy and the coevolution of cetacean retinal opsins', PloS One doi.org/10.1371/journal.pgen.1003432
acorn barnacles ... all over her body, de rigueur *for humpback whales, had five razor-sharp calcareous plates*
Jones, D and G Morgan (1994) *A Field Guide to Crustaceans of Australian Waters*, Reed New Holland, second edition, p. 224.

On the belly of a whale
Humpback whales, like all cetaceans ... possess a vibrant flora of bacteria and ... viruses. Directly breathing in the exhalation is not recommended.
Nelson, TM, A Apprill, J Mann, TL Rogers and MV Brown (2015) 'The marine mammal microbiome: current knowledge and future directions', *Microbiology Australia*, March (2015): 8-13.
the health of populations can be assessed from the microbes of an individual cetacean's microbiome
Nelson et al (2015).

Protecting the Kimberley
the north-eastern Indian Ocean via the Indo-Pacific through-flow, which then mixes with the Indian Ocean.
Fitzsimons, J and G Wescott (2016) *Big, Bold and Blue*, CSIRO Publishing Australia, p. 416.
The Kimberley's vast inland area of some 423 000 square kilometres is spectacularly beautiful and so rugged it remains a true wilderness region
Wilson, I (2006) *Lost World of the Kimberley*, Allen and Unwin, Sydney, p. 315.
the natural processes forming its geology are estimated to be 1.8 billion years old
Gueho R (2007) *Rhythms of the Kimberley*, Fremantle Press, North Fremantle, p. 208
Dinosaur footprints on the beach at Gantheaume Point, Broome, and at James Price Point date back 130 million years
ABC News: World's biggest dinosaur found in north-western WA: www.abc.net.au/news/2017-03-27/world-biggest-dinosaur-footprint-found-north-western-wa/8391098
remnants of stromatolites and blue green algae discovered at Marble Bar are 3500 million years old
ABC News: Scientists discover earliest signs of life in Pilbara: www.abc.net.au/news/2013-11-13/scientists-discover-earliest-signs-of-life-on-earth-in-pilbara/5088190
The idea of reserving marine areas for environmental protection and resource management began in the 1960s
Fitzsimons and Wescott (2016), p. 416.
our discovery that Camden Sound is a principal breeding area for the humpback whale
Fitzsimons and Wescott (2016), p. 416.
while filming a National Geographic documentary, Birthplace of the Giants
Sea Dog TV International (2015) *Birthplace of the Giants*, Nat Geo Wild documentary, 58 mins.

From the first radiation, cetaceans are represented in the fossil record 45 million to 53 million years ago in the Eocene era

Jefferson, TA, MA Webber, and RL Pitman (2015) *Marine Mammals of the World*, Academic Press, London, San Diego, Waltham, Oxford, p. 608.

The orphan

Battle scars are gained by males competing for females on the breeding grounds during the warm, tropical winter months

Chu, K and S Nierkirk (1988) 'Dorsal fin scars as indicators of age, sex and social status in humpback whales (*Megaptera novaeangliae*)', *Canadian Journal of Zoology* 66(2): 416–420.

Male humpback whales migrate to the calving grounds to mate as often as possible with available females

Darling, JD (2001), p. 61.

One was slightly larger

Clapham, PJ (1996) 'The social and reproductive biology of humpback whales: an ecological perspective', *Marine Mammal Review* (21)1: 27–49.

and less scarred

Chu, K and S Nierkirk (1988)

sperm whale calves are cared for and nursed by their mothers and non-mothers in alloparental care

Whitehead, H (1995) 'Babysitting, dive synchrony and indications of alloparental care in sperm whales', *Behav Ecol Sociobiol* (1996) 38: 237–244.

dolphins also engage in care of others' calves

Riedman, ML (1982) 'The evolution of alloparental care and adoption in mammals and birds', *The Quarterly Review of Biology* 57(4): 405–435.

Family ties are greater and totally different from those of baleen whales such as humpback whales

Whitehead, H and L Rendell (2015) *The Cultural Lives of Whales and Dolphins*, University of Chicago Press, Chicago, p. 417.

Humans nurse non-biological children as so-called wet-nurses in short-term care

Moran, L and J Gilad (2007) 'From folklore to scientific evidence: breast-feeding and wet-nursing in Islam and the case for non-puerperal lactation', *International Journal of Biomedical Science* 3(4): 251–257.

non-mothers have been able to lactate babies in long-term adoptive situations when the unique psychophysical trigger provides the let-down response

www.aboutkidshealth.ca/En/News/NewsAndFeatures/Pages/Breastfeeding-without-pregnancy.aspx.

This behaviour sparks the question: are humpback whales altruistic?

Pitman, RL, VB Deeke, CM Gabriele, M Srinivasan, N Black, J Denkinger, JM Durban, EA Mathews, DR Matkin, JL Neilson, A Schulman-Janiger, D Shearwater, P Stap and R Ternullo (2016) 'Humpback whales interfering when mammal-eating killer whales attack other species: mobbing behaviour and interspecific altruism?', *Marine Mammal Science* 33(1): 7–58.

Close pass

Seven cetacean species are reported to mourn their dead
Reggerte, MAL, F Alves, C Nicolau, L Freitas, D Cagnazzi, RW Baird and P Galli (2016) 'Nurturant behaviour toward dead conspecifics in free-ranging mammals: new records for odontocetes and a general review', *Journal of Mammology* 97(5): 1428–1434.
researchers at Curtin University and Murdoch University had named her Tupac
River Guardians (2017) *Finbook*, Government of Western Australia. www.riverguardians.com/projects/dolphin-watch/identifying-dolphins/.

Sunset thrash

once it's a yearling a calf separates from its mother
Darling, JD (2001), p. 61.
a team of dedicated researchers have had the perhaps unenviable task of towing dead beached whale carcasses offshore for sinking and monitoring.
Smith, CR, AG Glover, T Trevde, ND Higgs and DJ Amon (2015) 'Whale fall ecosystems: recent insights into ecology, paleoecology and evolution', *Annual Review of Marine Science* 7: 571–596.

Singers

Humpback whales are renowned for producing the most complex songs, not just sounds, in the animal kingdom
Payne RS and S McVay (1971) 'Songs of humpback whales', *Science* 173 (3997): 585–597.
complex and diverse songs that comprise phrases, sub-phrases and units (the smallest piece) in songs that last from 10 to 40 or so minutes
Payne and McVay (1971).
The evidence of such a rapid change in the song was deemed a 'cultural revolution' rather than a 'cultural evolution'
Noad, MJ, DH Cato, MM Bryden, M-NM Jenner and KSC Jenner (2000) 'Cultural revolution in whale songs', *Nature* 408, 537.

Satellite tags

Several functioning tags revealed a number of interesting details of their southern migration
Double, MC, N Gales, KCS Jenner and MN Jenner (2010) 'Satellite tracking of south-bound female humpback whales in the Kimberley region of Western Australia'. Final Report to AMMC, p. 30.

BLUE WHALES

Within the blue

like all cetaceans, replace their outer skin cells frequently, making collection a benign and potenitally useful tool
Gendron, D and SL Mesnick (2001) 'Sloughed skin: a method for the systematic collection of tissue samples from Baja California blue whales', *Journal of Cetacean Research and Management* 3(10): 77–79.

useful genetic material slough skin contained due to the lower quantity of live DNA in these dead external cells

Amos, W and AR Hoelzel (1991) 'Long-term preservation of whale skin for DNA analysis', *Report of the International Whaling Commission* (Special Issue) 13: 99–103.

there are two alternative strategies for baleen whales, one that results in fat whales and one that results in streamlined whales

Whitehead, H and L Rendell (2015) *The Cultural Lives of Whales and Dolphins*, University of Chicago Press, Chicago, p. 417.

dead krill specimens were compared with those found in the faecal material and the conclusion was made that these pygmy blue whales were indeed feeding in the deep waters of the Perth Canyon

McCauley, RD, JL Bannister, C Burton, C Jenner, S Rennie and CS Kent (2004) 'Western Australian Exercise Area Blue Whale Project. Final Summary Report Milestone 6: Sept 2004', *CMST Report* R2004-29, project350, p. 71.

as many as 27 other krill species that have now been identified in the Perth Canyon

Sutton, AL (2015) *Krill in the Leeuwin Current: influence of oceanography and contribution to Indian Ocean zoogeography*, PhD thesis, Murdoch University, p. 189.

investigate the population status of blue whales south of Australia, where historically the Russian whalers had taken thousands.

Branch, TA, KM Stafford, DM Palacios, C Allison, JL Bannister, CLK Burton, E Cabrera, CA Carlson, B Galleti Vernazzani, PC Gill, R Hucke-Gaete, KCS Jenner, M-NM Jenner, K Matsuoka, YA Mikhalev et al (2007) 'Past and present distribution, densities and movements of blue whales *Balaenoptera musculus* in the Southern Hemisphere and northern Indian Ocean', *Mammal Review* 37: 116–175 doi: 10.1111/j.1365-2907.2007.00106.

reported whales as 'like-blue' and 'blue whale' with 'up to two per day recorded in December 1995 in a small area some 25 nautical miles off Rottnest Island, in the area of the Perth Canyon …'

Bannister, JL (2008) *Great Whales*, CSIRO Publishing, Collingwood, p. 142

During the whaling era, blue whales were highly prized and targeted most frequently, since the yield per unit of effort was greatest.

Dakin, WJ (1934) *Whalemen Adventurers*, Angus and Robertson, first edition 1934, revised edition (1963) Sirius Books p. 285.

One blue whale could yield at least twice as much oil as a humpback whale (on average 70 to 80 barrels versus 35 to 40 barrels).

Dakin, WJ (1934), p. 285.

this female Antarctic blue whale was killed in February 1928, and was recorded as measuring 30.5 metres and weighing 160 tonnes.

Payne, R (1995) *Among Whales*, Scribner, p. 431

biopsy samples for laboratory analysis by Dr Bob Brownell in California, who confirmed that these were pygmy blue whales

LeDuc, RA, A Dizon, M Goto, L Pastene, H Kato, S Nishiwaki, C LeDuc and R Brommell (2007) 'Patterns of genetic variation in Southern

Hemisphere blue whales and the use of assignment test to detect mixing on the feeding grounds', *Journal of Cetacean Research and Management* 9: 73–80.

However, their heads are relatively larger, giving them a 'tadpole' shape in contrast to the 'torpedo' shape of their larger relatives.

Bannister (2008), p. 142.

Don't we know you?

Using six 'capture' sessions (2000 to 2005) and the 208 individuals photographed from the 271 sightings, our mark-recapture model produced a population size estimate of 712 to 1754 individuals.

Jenner, KCS, M-NM Jenner, C Burton, V Sturrock, CP Salgado Kent, M Morrice, C Attard, L Moller and MC Double (2009) 'Mark-recapture analysis of pygmy blue whales from the Perth Canyon, Western Australia 2000–2005', *Report to International Whaling Commission* SC/60/SH16.

Saving technology

they spent about two weeks here, diving to feed at depths of 400 to 500 metres along the shoulders of the deep-water trench.

McCauley, RD, JL Bannister, C Burton, C Jenner, S Rennie and CS Kent (2004) 'Western Australian Exercise Area Blue Whale Project. Final Summary Report Milestone 6: Sept 2004', *CMST Report* R2004-29, project350, p. 71.

a pygmy blue whale tagged in the Perth Canyon ... revealed it had travelled beyond the Western Australian coastline into the Banda Sea in Indonesia, a journey of approximately 3000 nautical miles.

Double, MC, V Andrews-Goff, KCS Jenner, M-NM Jenner, SM Laverick, TA Branch and NJ Gales (2014) 'Migratory movements of pygmy blue whales (*Balaenoptera musculus brevicauda*) between Australia and Indonesia as revealed by satellite telemetry', *PloS One* 9(4); e93578.

possibly a humpback whale with lots of white on its flanks (a Type 1 or Type 2 pigmentation)

Kaufman, GD, M Smultea and PH Forrestall (1987) 'Use of lateral body pigmentation patterns for photographic identification of east Australian (Area V) humpback whales', *Cetus* 7: 5–13.

A week in the life of a pygmy blue whale

Over the course of the nearly eight days of attachment, this pygmy blue whale made 1677 dives, the average being to a depth of 13 metres

Owen, K, CS Jenner, M-NM Jenner and RD Andrews (2016) 'A week in the life of a pygmy blue whale: migratory dive depth overlaps with large vessel drafts', *Animal Biotelemetry* 4:17. doi.10.1186/s40317-016-0109-4

Eye to eye

As the largest animal, including the biggest dinosaur, that has ever lived on earth you could afford to be gentle

Payne, R (1995), p. 431.

Ancient tracks

As the whale travels through the shifting, changing seas, a map — unlike anything used on land — is learned.

Warshall, P (1974) 'The way of the whales' in J McIntyre (ed.), *Mind in the Waters*, Charles Scribner's Sons, New York, pp. 110–131.

Blue whale facts

A blue whale's heart weighs two tons

Payne, R (1995), p. 431.

Whaling then and now

Prior to commercial whaling, it has been calculated that the Antarctic blue whale population numbered 239 000

Branch, TA, KM Stafford, DM Palacios, C Allison, JL Bannister, CLK Burton, E Cabrera, CA Carlson, B Galleti Vernazzani, PC Gill, R Hucke-Gaete, KCS Jenner, M-NM Jenner, K Matsuoka, YA Mikhalev et al (2007) 'Past and present distribution, densities and movements of blue whales *Balaenoptera musculus* in the Southern Hemisphere and northern Indian Ocean', *Mammal Review* 37: 116–175 doi: 10.1111/j.1365-2907.2007.00106.

A study conducted in 1996 south of Madagascar by the late Peter Best and colleagues estimated the population to be 424

Best, PB, RA Rademeyer, C Burton, D Ljungblad, K Sekiguchi, H Shimada, D Thiele, D Reeb and DS Butterworth (2003) 'The abundance of blue whales on the Madagascar Plateau, December 1996', *Journal of Cetacean Research and Management* 5(3): 253–260.

our collaborative Perth Canyon research using aerial surveys estimated the Perth Canyon population at 30

McCauley, RD, JL Bannister, C Burton, C Jenner, S Rennie and CS Kent (2004) 'Western Australian Exercise Area Blue Whale Project. Final Summary Report Milestone 6: Sept 2004', *CMST Report* R2004-29, project350, p71.

Using photo-ID, a closed time-dependent model, we estimated population size as 712 to 1754 individuals.

Jenner, KCS, M-NM Jenner, C Burton, V Sturrock, CP Salgado Kent, M Morrice, C Attard, L Moller and MC Double (2009) 'Mark-recapture analysis of pygmy blue whales from the Perth Canyon, Western Australia 2000–2005', *Report to International Whaling Commission* SC/60/SH16.

with photo-ID, genetic matches and sat-tagging pygmy blue whales

Attard, CR, LB Behregaray, C Jenner, P Gill, M Jenner et al. (2010) 'Genetic diversity and structure of blue whales (*Balaenoptera musculus*) in Australian feeding aggregations', Conserv Genet II: 2437–2441. Double, MC, V Andrews-Goff, KCS Jenner, MN Jenner, SM Laverick, TA Branch and NJ Gales (2014), 'Migratory movements of pygmy blue whales (*Balaenoptera musculus brevicauda*) between Australia and Indonesia as revealed by satellite telemetry', *PlosOne* April 2009, 2014, http://doi.org/10.1371/journal.pone.0093578.

DOLPHINS

Dolphins on the bow!

cetacean researcher Dr Deb Thiele and her team revealed that their most-used habitat was indeed in the busiest part of the Kimberley, Roebuck Bay.

Thiele, D (2010) *Collision Course: Snubfin Dolphin Injuries in Roebuck Bay, a report prepared for WWF Australia.*

*Genetically different from other species, these dolphins have been correctly identified as Australian snubfin dolphins (*Orcaella heinsohni*)*

Beasley, I, K Robertson and P Arnold (2005) 'Description of a new dolphin, the Australian snubfin dolphin (*Orcaella Heinsohni* sp. N C *Cetacea, Delphinidae*)', *Marine Mammal Science* 21(3): 365–400.

Diamond dolphins

Researchers from Murdoch University have observed over 45 octopus handling cases of bottlenose dolphins preparing octopus for consumption,

Sprogis, KR, HC Raudino, D Hocking and L Bejder (2017) 'Complex handling of octopus by bottlenose dolphins (*Tursiops aduncus*)', *Marine Mammal Science*, doi:10.1111/mms.12405.

Gilligan, an adult male observed since 2007, was found dead on Stratham Beach near Bunbury in Western Australia in August 2015 with an octopus protruding from his mouth.

Stephens, N, P Duignan, J Symons, C Holyoake, L Bejder and K Warren (2017) 'Death by octopus (*Macroctopus maorum*): laryngeal luxation and asphyxiation in a bottlenose dolphin (*Tursiops aduncus*)', *Marine Mammal Science* doi.10.1111/mms.12420.

Common ground

understand the genetic diversity of short-beaked common dolphins across the south coast of Australia, where they are thought to be at risk from local fishing operations

Bilgmann, K, LM Moller, RG Harcourt, R Gales and LB Beheregaray (2008) 'Common dolphins subject to fisheries impacts in Southern Australia are genetically differentiated: implications for conservation', *Animal Conservation* 11(6): 518–528.

Killer whale surprise

Sadly, the southern resident pods, types J, K and L (fish eaters) studied carefully by Ken Balcomb III and the late Prentice Bloedel II at Orca Survey, are under grave threat

Wasser, SK, JI Lundin, K Ayrels, E Seely, D Giles, K Balcomb, J Hempelmann, K Parsons and R Booth (2017) 'Population growth is limited by nutritional impacts on pregnancy success in endangered southern resident killer whales (*Orcinus orca*)', *PloS One*, 2017; 12(6): e0179824. Doi.10.1371/journal.pone.0179824.

possibly due to several networks, such as Killer Whales Australia

Killer Whales Australia: www.facebook.com/killerwhalesaustralia/

Project ORCA: www.projectorca.com.au/ and www.facebook.com/orcatalkoz/

Since 2006, these visits have been for longer periods of time as these animals are actively predating on neonate humpback whales

Pitman, RL, JA Totterdell, H Fearnbach, LT Balance, JW Durban and H Kemps (2014) 'Whale killers: prevalence and ecological implications of killer whale predation on humpback whale calves off Western Australia', *Marine Mammal Science* 31(2): 629–657.

killer whales appear every summer and early autumn, congregating in pods of 50 to 100 in the deep water beyond the 200-metre continental shelf at the Bremer Canyon

cmst.curtin.edu.au/wp-content/uploads/sites/4/2017/04/Killer-Whales-Bremer-Sub-Basin-Catalogue.1st-Edition.-Wellard-2017.-Project-ORCA.pdf

photographically as close to two dozen pods of just over 80 individually photo-ID'd animals

Wellard, R, K Lightbody, L Fouda, M Blewitt, D Riggs and C Erbe (2016), Killer Whale (*Orcinus orca*) predation on beaked whales (*Mesoplodon* spp) in the Bremer sub-basin, Western Australia, *PlosOne* 11(12): e016670.doi.org/10.1371/journal.pone.0166670.

Occasionally, beaked whales are preyed upon by the killer whales, which have been documented photographically as close to two dozen pods of just over 80 individually photo-ID'd animals

cmst.curtin.edu.au/wp-content/uploads/sites/4/2017/04/Killer-Whales-Bremer-Sub-Basin-Catalogue.1st-Edition.-Wellard-2017.-Project-ORCA.pdf

Studies currently being conducted on the genetics of the killer whales indicate that these whales share some similar genetic material with the transient (or marine mammal eating) Bigg's whales in the Pacific Northwest.

Wellard, R, K Lightbody, L Fouda, M Blewitt, D Riggs and C Erbe (2016) 'Killer Whale (*Orcinus orca*) predation on beaked whales (*Mesoplodon spp*) in the Bremer Sub-Basin, Western Australia', *PloS One* 11(12): e0166670. doi.org/10.1371/journal.pone.0166670.

MINKE WHALES

Minke whale magic

This pilot study with four satellite LIMPETs (Low Impact Minimal Profile Electronic Tags)

Cooke, SJ, SG Hinch, M Wikelski, RD Andrews, LJ Kuchel, TG Wolcott and PJ Butler (2004) 'Biotelemetry: a mechanistic approach to ecology', *Trends in Ecology and Evolution* 19(6): 334–343.

The regulations regarding the swim-with permits, such as the one held by JR for dwarf minke whales on the Great Barrier Reef, stipulate that all swimmers must hold the line at all times.

www.minkewhaleproject.org/

The 'ba-da-da-doing' sound of these animals, also known as the 'Star Wars *whale calls'*

Gedamke, J, DP Costa and A Dunstan (2001) 'Localisation and visual verification of a complex minke whale vocalisation', *Journal of the Acoustical Society of America* 109.3038(2001). Doi.dx.doi.org/10.112y1.1371763.

captured on the sonobuoys (underwater listening devices)
Holler, RA (2014) 'The evolution of the sonobuoy from World War II to the Cold War', *US Navy Journal of Underwater Acoustics*, January 2014, p. 323, pdf.
processed through sophisticated equipment such as SDR (Software Defined Radio)
Rudra, A and A Bose (2006) 'High speed ADC combines with FPGA to enable single-slot 46-50 SDR solutions', www.rfdesign.com defenseelectronicsmag.com/sitefiles/defenseelectronicsmag.com/files/archive/rfdesign.com/mag/604RFDF4.pdf.

Minke mischief
whale sightings in the Great Barrier Reef from April to September each year, but 90 per cent of the sightings are made in June and July
Arnold, P (1998) 'Occurrence of dwarf minke whales (*Balaenoptera acutorostrata*) on the northern Great Barrier Reef, Australia', *Report for the International Whaling Commission* 47: 419–424.
www.minkewhaleproject.org/.
The data collected over the previous 18 years showed that the dwarf minke whales ranged in size from 3.7 metres to 7.0 metres in length
www.minkewhaleproject.org/.
Most of the whales visited during the peak winter months, but where did they go the other 10 months of the year?
www.minkewhaleproject.org/.
and began to spy hop: its head broke the surface and its body was vertical in the water as we faced each other
WDC Ethogram *Humpback Whale Behavior 2013* (pdf): whalesenseblog.files.wordpress.com/2013/04/ethogram_mn.pdf.

Saving whales
The data revealed by the tags on the dwarf minke whales was revolutionary
http://www.abc.net.au/news/2013-08-15/study-to-identify-threats-to-reef-minke-whales/4888908.

LISTENING IS THE NEW LOOKING
Bringing sights and sounds together
'A ship towing an array of hydrophones can effectively hear whales in a strip 150 miles wide — an improvement of seventy-five times'
Payne, R (1995), p. 431.
'have been observed responding to ships by slowly sinking below the surface'
Jefferson, TA, MA Webber, and RL Pitman (2015) *Marine Mammals of the World*, Academic Press, London, San Diego, Waltham, Oxford, p. 608.
One day as we were working offshore, our colleague Chris 'Daffy' Donald was monitoring a sonobuoy
The Defence team used what they called out-of-life sonobuoys for all of the experiences described in this chapter. These sonobuoys had reached the end of their shelf life and would otherwise have been crushed by a contractor to prevent unauthorised use. Defence indicated that the WAAP

in transit not only saved on disposal costs but also gave them the multiple benefits of constantly evolving the sonars on board and learning more about operating environments, particularly how they were trending. They also seemed pleased to be using the buoys rather than simply destroying them.

Dolphins and dynamite

The fishermen were prosecuted by the Australian authorities

Australian Fisheries Management Authority 'Illegal blast fishers convicted and boats destroyed' 26 October 2012: www.afma.gov.au/illegal-blast-fishers-convicted-and-boats-destroyed/.

The future is in listening

cetacean species represented by acoustic spectrograms are new recordings of pygmy killer whales, striped dolphins, Omura's whales and Fraser's dolphins

Erbe, C, R Dunlop, C Jenner, M Jenner, R McCauley, I Parnum, M Parsons, T Rogers and C Salgado Kent (2017) 'Review of underwater and in-air sounds emitted by Australian marine mammals' (in press).

The frequency of a call represents the number of times per second compressions and rarefactions of the water molecules occur as the sound passes through the water

Payne, R (1995), p. 431.

the frequency of the call produced by pygmy blue whales has decreased by one hertz over nine years

Gavrilov, AN and RD McCauley (2013) 'Acoustic detection and long-term monitoring of pygmy blue whales over the continental slope in south-west Australia', *Journal of Acoustical Society* AM 134(3): 2505–2513.

COOL WATER: ANTARCTICA

Twenty years in the making

krill, Euphausia superba *— the cornerstone species of the rich Antarctic ecosystem*

Australian Antarctic Division: 'Krill': www.antarctica.gov.au/science/conservation-and-management-research/southern-ocean-fisheries/krill.

Another theory proposes seeking calm water as the driver for seasonal migrations

Avgar, T, G Street and JM Fryxell (2014), 'On the captive benefits of mammalian migration', *Canadian Journal of Zoology*, 92(6): 481–490, http://doi.org/10.1139/cjz-2013-0076.

acoustic recordings show that Antarctic blue whales, which spend the southern hemisphere summer in the Antarctic, also occasionally visit the Perth Canyon in winter.

McCauley, RD, JL Bannister, C Burton, C Jenner, S Rennie and CS Kent (2004) 'Western Australian Exercise Area Blue Whale Project. Final Summary Report Milestone 6: Sept 2004', *CMST Report* R2004-29, project350, p. 71.

During whaling, the population of Antarctic blue whales was reduced from some 300 000 to only around 6000 animals. Pygmy blue whales, usually found north of latitude 40 degrees south, fell from around 10 000 to around 6000.

Branch, TA, KM Stafford, DM Palacios, C Allison, JL Bannister, CLK Burton, E Cabrera, CA Carlson, B Galleti Vernazzani, PC Gill, R Hucke-Gaete,

KCS Jenner, M-NM Jenner, K Matsuoka, YA Mikhalev et al (2007) 'Past and present distribution, densities and movements of blue whales *Balaenoptera musculus* in the Southern Hemisphere and northern Indian Ocean', *Mammal Review* 37: 116–175 doi: 10.1111/j.1365-2907.2007.00106.

the Antarctic Circumpolar Current, the world's fastest flowing current, which travels eastward at 153 million cubic metres per second

Lonely Planet (2008) *Antarctica,* Lonely Planet Publications, p. 380.

A plethora of wildlife — seabirds, seals, porpoises, dolphins and whales — rely on summer krill patches, the basis of the Antarctic ecosystem

Griffiths, HJ (2010) 'Antarctic marine biodiversity — what do we know about the distribution of life in the Southern Ocean?' *PloS One* 2010: 5(8):e11683. doi:10.1371/journal.pone.0011683.

Tracking Tiny

Tiny Lander's tail fluke was very distinctive with an easily recognisable black 'butterfly' or # 55 fluke pattern on the underside.

spo.nmfs.noaa.gov/mfr653/mfr6532.pdf

Mizroch SA and AD Harkness (2003) 'A test of computer-assisted matching using the north Pacific humpback whale (*Megaptera novaeangliae*) tail flukes photographic collection', *Marine Fisheries Review*, 65(3): 25–37.

Hungry whales and fussy eaters

blue whales feed singularly and opportunistically, even apparently 'choosing' whether to expend energy in the process of gaining energy in relation to the size of krill mass presented

Video from scientists at Oregon State University taken in the Southern Ocean off New Zealand: phys.org/news/2017-04-video-blue-whales-strategy.html.

Sperm whales

Researchers Hal Whitehead and Luke Rendell have dedicated their lives to unravelling the mysteries of these beautiful animals, including detailing their extraordinary society

Whitehead, H and L Rendell (2015) *The Cultural Lives of Whales and Dolphins*, University of Chicago Press, Chicago, p. 417.

Sperm whales are the most sexually dimorphic cetaceans, males weighing almost three times as much as the females

Jefferson, TA, MA Webber, and RL Pitman (2015) *Marine Mammals of the World*, Academic Press, London, San Diego, Waltham, Oxford, p. 608.

FURTHER READING

Conefrey, M (2005) *A Teacup in a Storm*, Collins, London, p. 240.

Curtin University Centre for Marine Science and Technology: cmst.curtin.edu.au/.

Dawbin, WH (1966) 'The seasonal migratory cycle of humpback whales', in KS Norris (ed.) *Whales, dolphins and porpoises*, University of California Press, Berkeley and Los Angeles, pp. 145–170.

Day, D (1992) *The Whale War*, HarperCollins, Glasgow, p. 206.

Horden, M (1997) *King of the Australian Coast*, Melbourne University Press, p. 44.

Lalang-garram/Camden Sound Marine Park management plan 73, 2013–2023, http://www.dpaw.wa.gov.au/images/documents/parks/management-plans/2012041_Lalang-garram_Camden_Sound_Marine_Park_MP_2013-2023_WEB.pdf.

McCauley, RD, KCS Jenner, M-NM Jenner, KA McCabe and J Murdoch (1998) 'The response of humpback whales (*Megaptera novaeangliae*) to offshore seismic noise: preliminary results of observations about a working seismic vessel and experimental exposures', *APPEA Journal* 1998: 692–707.

Melville, H (1851) *Moby Dick*, p. 630.

The Minke Whale Project: www.minkewhaleproject.org/.

Murdoch University Cetacean Research Unit: mucru.org/.

Neiwert, D (2015) *Of Orcas and Men*, The Overlook Press, Peter Mayer Publishers, Inc, New York, p. 305.

Pash, C (2008) *The Last Whale*, Fremantle Press, p. 218.

Peterson, B and L Hogan (2002) *Sightings*, National Geographic Society, Washington DC, p. 286.

Preston, D and M (2004) *A Pirate of Exquisite Mind*, Doubleday, Great Britain, p. 512.

Project ORCA: www.projectorca.com.au/.
Safina, C (1998) *Song for the Blue Ocean: Encounters Along the World's Coasts and Beneath the Seas*, Holt Paperbacks, p. 480.
Safina, C (2015) *Beyond Words: What Animals Think and Feel*, Picador, p. 461.

ACKNOWLEDGMENTS

During my three decades of whale research in the US and Australia, many people have been involved, but my first thank you must go to Curt, my partner in crime for this adventure. Hours of his moustache-twirling while 'thinking like a whale' has formulated the process of our whale research explorations along the windiest coast in the world. Scientific research demands repetition, but with learning one must adjust, and for these reasons our research expeditions have taken changes in tack, camping on offshore desolate islands and even turning our hands to boat-building in order to find the northern destination for humpback whales in the Kimberley. This has been a great ride so far and promises to continue to be so. Thanks, Curt.

We are immensely grateful to the late Prentice Bloedel II, who kindly offered a loan (all-be-it written on the back of a used envelope and signed on the bonnet of a car!), which enabled us to begin research in the Dampier Archipelago. Without his backing and generosity, none of these experiences could have occurred.

Thanks go to our dear friend John Bannister, former director of the Western Australian Museum, who efficiently assisted with logistics on many levels, particularly with access to the Conservation and Land Management research station on Enderby Island in the Dampier Archipelago, and with darkroom/office facilities at the Western Australian Museum in the early 1990s.

Dave Mel and Hugh Chevis ably provided logistics for us to use the Conservation and Land Management research station and storage shed.

We appreciate Douglas Elford, photographer at the Western Australian Museum, for his friendship, expert advice on photography techniques and detailed Filemaker-Pro knowledge for building our WA Humpback Whale Catalogue (our computer-based archival and matching system).

To David and Michelle Davenport, with whom we have forged a strong family friendship, we have appreciated your support in a multitude of ways, thank you.

Vanessa Sturrock ably and enthusiastically assisted us on remote Enderby Island in 1992 and 1993 as a young petro-geology graduate, and even helped build *WhaleSong*; Vanessa was also there for the 2004 pygmy blue whale season at Rottnest Island on board *WhaleSong*, and aboard *WhaleSong II* for the 2007 humpback whale season.

John and Lyn Lally at the Pilbara Camp School were formative in getting us to the Kimberley, especially in convincing us to build a boat, and providing multiple $5 trade secrets during *WhaleSong*'s construction. We will always remain grateful for your expert boat-building advice.

Chris Lally, builder-extraordinaire, expertly hurried the build process, then with a steady hand, hilariously poetic ship's log entries and a very wicked sense of humour captained *WhaleSong* on her first two years of exploring in the Kimberley.

Leon Janssen beautifully prepared the metalwork on *WhaleSong* and we appreciate his keen eye for detail — and for giving up his Sundays for almost a year in exchange for lunch.

Mandie Thompson and Renee Mazeroll, both Earthwatch volunteers who personally raised sponsorship funds to help build *WhaleSong* are remembered fondly and thanked appreciatively.

Our first mate on *WhaleSong*, Katie McCabe, a seasoned world sailor who had 10 000 nautical miles under her belt when she joined us, provided three years of pure dedication to the craft of sailing, detailed research and producing yummy culinary treats with a generous side order of laughs.

Kaye Grubb is to be thanked for understanding our intense interest in the Kimberley, Charlie Grubb from Coastwatch for his monthly envelopes of hand-written whale sightings and, in a nice twist, Nicole Grubb (their daughter) for her dedicated transfer of her dad's sightings plotted on a nautical chart to a computer chart almost 15 years later.

We gratefully acknowledge the enduring friendship and assistance of Dale and Liz Peterson since meeting them in 1995. Dale has sailed thousands and thousands of miles on all three vessels, *WhaleSong*, *WhaleSong II* and *Whale*

Song as our first mate and, incredibly, found our current *Whale Song* while reading a magazine.

Our marine crew, who keep all our vessels functioning, have been integral to this whole operation. They include Resty Adenir, Kadin Anketell-Walker, Fred Gonio, Elizabeth Howell, Simon Kenion, Ross McLaren-Nicole, Julie Murdoch, Wendy Pentland, Dale Peterson, Dave Porter, Pete Sandison, Nick Sambrooks, Warren Sharpe, Glynn Thoman, Ken Waller, Brian Wallis and Sam Wright.

The science crew (research associates and research assistants), comprising more than 50 personnel, have been essential for documenting all the research findings, data write-up and analysis for shipboard surveys and aerial surveys. To this willing and able young crew, we wholeheartedly thank you and acknowledge we couldn't have done it without you.

We thank our hard-working crew during the Earthwatch seasons (a dozen dedicated souls) and we appreciate the very keen and excited Earthwatch volunteers, numbering 120, who ably assisted us during the research seasons of 1992, 1993 and 1995 to 1998.

Wayne and Pam Osborn, we appreciate your dedication to excellent photography, including contributing to the book cover, our research catalogues and scientific endeavours. You two have been thoroughly bitten by the whale bug!

Rob McCauley has been an integral scientist in many Centre for Whale Research projects over the last 22 years. We appreciate your family friendship, calmness and focused nature.

To Chandra Salgado Kent, we are grateful for your kindness, attention to detail and expert statistical analysis skills — you are second to none!

To our scientific colleagues Russell Andrews, Catherine Attard, John Bannister, Lynnath Beckley, Lars Bejder, Kerstin Bilgmann, Chris Burton, Doug Cato, Mike Double, Christine Erbe, Stu Field, Pete Gill, Nick Gales, Alexander 'Sasha' Gravilov, Megan Huggett, Benjamin Kahn, Rob McCauley, Jessica Meeuwig, Luciana Moller, Holly Raudino, Chandra Salgado Kent, Michele Thums and Kelly Waples, we thank you for many great collaborative studies.

To our past students, Dr Muriel Brasseur, Dr Philippe Bouchet, Dr Janelle Braithewaite and Gabrielle Cummins (BSc. Hons) — congratulations! And to our current PhD candidates Carrie Skorcz and Tamara Al Hashimi — go well!

For our new role supporting the Royal Australian Navy envisaged by Christopher 'Daffy' Donald: we thank you, Daffy, for being so radically visionary, unwaveringly determined and guiding us for the first two years while developing *Whale Song*'s unique capabilities. As well, she has proven to be an excellent platform for collecting high-calibre acoustic scientific data, spectacularly returning her to original listening role for cetacean research.

To Wayne Bennett, who has picked up Daffy's enthusiasm for a challenge and his team at the Department of Defence, we appreciate your steady hands in these exciting and uncharted waters.

We thank the personnel from the Australian Department of Defence and SUBFOR, L3 Oceania and Sonartech

Atlas, our regular acoustic crew aboard *Whale Song*, for being such great work mates.

With so many names to acknowledge, please forgive any ommissions, you know we have appreciated you.

Many companies, corporations and individuals have been involved in the almost 100 cetacean research projects conducted by the Centre for Whale Research since 1990. We are most grateful for the assistance of each one.

Funding and logistics sponsors

Wiltrading (WA) Pty Ltd has been a key sponsor of the Centre for Whale Research since 1990, Advantage Elvstrom North Sails, Allyacht Spars, API Pty Ltd, Australian Geographic Society, Baileys Marine Fuels, BHP Billiton Petroleum, Bristow Helicopters, Center for Whale Research (USA), Cetacean Society International, Charters and Co., Coastwatch, Crushing and Mining Equipment, Curtin University, Dampier Port Authority, Department of Conservation and Land Management, Department of Customs and Immigration, Department of Defence, Earthwatch and EarthCorps volunteers, Edith Cowan University, Environment Australia (Department of Environment and Heritage), Hampton Harbour Boat and Sailing Club, Icom America, Ilford Australasia, INPEX Ltd, International Paints (who have supported the Centre for Whale Research since 1995), International Whaling Commission, John F. Long Foundation, Kimberley Camp School, Leeuwin Sail Training Association, Maurice Buckley and Poole and Permanent Trustee Corporation, Morgan and Son, Murdoch University, Noble and Sons, North Port Boat Lifters, Novurania, Pilbara Camp School, Plantagenet

Press, Propsero Productions, ROC Oil, Ronstan Australia, Rottnest Island Authority, Royal Perth Yacht Club, Sea Dog TV International, Shell Development Australia, Storyteller Productions, University of Western Australia, Weiland Agencies, Western Australian Museum, Western Australian Petroleum, Western Mining Corporation, Woodside Energy, Yacht Grot, Yamaha Motors and Perth Advocates for the Earth Inc.

Private individual funding
Pam Bebreen, the late Prentice Bloedel II, Micheal Caplehorn, David and Michelle Davenport, Ted and Shirley Davis, Debbie Dorand, Kaye and Charlie Grubb, Ken and Irma Jenner, Roger Kempe, Elsie Lennell, Tot Long, the late Liz McLaren-Nicole, Jim Nahmens, Chris Nicastro, Wayne and Pam Osborn, Gaynell Schenck, the late Newton Sinclair and the late Frances Sinclair, the late Selby Smith and the late Emma Smith, Vanessa Sturrock, Ron and Cherrie Taylor, John Walker, Guy and Kate Wright and Martin and Fiona Webster.

Equipment donation/discount
Advantage Elvstrom South Sails, AFPT Foam, Allyacht Spars, AMI Sales, Arrow Electrical, Autoblast, Barrett, Battery Energy Power Solutions, Container Refrigeration, Farinosi and Sons, Fiocchi Munizioni SPA, Fremantle Hydraulics, Hamilton Engineering, Icom America, International Paints, Morgan Timbers, Noble and Sons, Prop Speed, Saint Gobain, speedcast, System III Australia, Yacht Grot, Yamaha Fremantle and Wiltrading Pty Ltd.

Private individual assistance
C Scott Baker, Ken Balcomb III, Francis Baronie, the late Prentice Bloedel II, Chris and Jess Bray, Andrew Davenport, David and Michelle Davenport, Leighton and Jodie DeBarros, Christopher 'Daffy' Donald, Debbie Dorand, John Garnett, Bruno Gicquel, Richard Holst, Wade and Robyn Hughes, Steve Katona, Simon Kenion, Craig Kitson, Megan Lally, Tot Long, the late Liz McLaren-Nicole, Ross McLaren-Nicole, Sally Mizroch, Jim Nahmens, Wayne and Pam Osborn, Dale and Liz Peterson, Anthea Porter, Peter Randall, Trevor Richards, Howard Rosenbaum, Jill Saint, Chandra Salgado Kent, Gaynell Schenck, Jenny Shaw, Luke and Jen Smith, the late Frits Steenhauer, Martin and Fiona Webster, Wayne and Barbie Williams, the late Grant Wilson and the late Helene Wilson. Thank you everyone, this work could not have been done without you!

o

I wish to acknowledge the 60 000 year presence of Indigenous Australians across this beautiful sunburnt, beach-fringed land. Curt and I are truly privileged to call this continent our home.

Thanks extend to Elspeth Menzies, Linda Funnell, Emma Hutchinson and all the team at NewSouth Publishing for expertly honing, maintaining grammatical pedantry and effective fine-tuning to present our story and the secrets of whales.

We acknowledge the Explorers Club, particularly as we carried EC Flag 69, and the Explorers Museum for bearing

TEM Pennant 2. We are honoured to join the ranks of such distinguished and intensely interesting global and scientific explorers.

Curt and I are forever grateful for the wonderful support of our families, my adventure-spirited mum (the late Liz McLaren-Nicole) who was our safety net and self-taught weather *aficionado* and my strong, seafaring brother Ross McLaren-Nicole, who fearlessly and ably sailed many of the trips north and south on *WhaleSong* and *WhaleSong II* with us. We are grateful to Ken and Irma Jenner, Curt's lovely parents, who have always been ready to provide encouragement regarding our boating capers and high seas adventures from non-seafaring country in western Canada. To Micah and Tasmin, our high-seas babies, thanks for making life extra exciting. To Skipper, my noisy white shadow – let's find some dolphins! And to the whales — well, thanks for all the fun!

Printed in the USA
CPSIA information can be obtained
at www.ICGtesting.com
JSHW070644090923
47974JS00013BA/100

9 781742 235547